Caves

The Natural Environment

Series Editors
Andrew Goudie and Heather Viles

This series will provide accessible and up-to-date accounts of the physical and natural environment in the past and in the present, and of the processes that operate upon it. The authors are leading scholars and researchers in their fields.

Published

Caves
Processes, Development and Management
David Gillieson

The Changing Earth
Rates of Geomorphological Processes
Andrew Goudie

Land Degradation
Laurence A. Lewis and Douglas L. Johnson

Oceanic Islands
Patrick D. Nunn

Humid Tropical Environments
*Alison J. Reading, Russell D. Thompson
and Andrew C. Millington*

Forthcoming

Rock Slopes
Robert Allison

Drainage Basin
Form, Process and Management
K. J. Gregory and D. E. Walling

Deep Sea Geomorphology
Peter Lonsdale

Holocene River Environments
Mark Macklin

Wetland Ecosystems
Edward Maltby

Arctic and Alpine Geomorphology
Lewis A. Owen, David J. Evans and Jim Hansom

Earth Surface Systems
Order, Complexity and Scale
Jonathan Phillips

Weathering
W. B. Whalley, B. J. Smith and J. P. McGreevy

Caves: Processes, Development and Management

David Gillieson

BLACKWELL
Publishers

The right of David Gillieson to be identified as author of this work has been asserted in accordance with the Copyright, Designs and Patents Act 1988.

First published 1996

2 4 6 8 10 9 7 5 3 1

Blackwell Publishers Ltd
108 Cowley Road
Oxford OX4 1JF
UK

Blackwell Publishers Inc
238 Main Street
Cambridge, Massachusetts 02142,
USA

British Library Cataloguing in Publication Data
A CIP catalogue record for this book is available from the British Library.

Library of Congress Cataloging-in-Publication Data
Gillieson, David S.
    Caves: processes, development, and management / David Gillieson.
        p.   cm. – (The natural environment)
    Includes bibliographical references and index.
    ISBN 0-631-17819-8. – ISBN 0-631-19175-5 (pbk.)
    1. Caves. I. Title. II. Series.
    GB601.G5   1996
    551.4'47–dc20
96-12123

CIP

Typeset in 10 on 11½ pt Sabon
by Best-set Typesetter Ltd., Hong Kong
Printed in Great Britain by T.J. Press Limited, Padstow, Cornwall

This book is printed on acid-free paper

Contents

Preface and Acknowledgements

When I try to imagine a faultless love
Or the life to come, what I hear is the murmur
Of underground streams, what I see is a limestone landscape.
                    W. H. Auden: *In Praise of Limestone*

This is an unashamedly antipodean view of the world of caves: although the caves of the fragments of Gondwanaland are not the deepest or the longest in the world, they are among the oldest and have formed under conditions markedly different from those of the higher latitudes of the northern hemisphere. The inheritance of the ice ages is expressed differently in the caves of Australia, where the climatic fluctuations have been between dry and wet rather than cold and wet (except in Tasmania). I have included a selection of the caves of the tropical world, in recognition of their great scientific interest, of their great beauty and of the small but energetic group of tropical karst scientists who have placed them firmly on the global karst scene. I have also provided an overview of cave management issues in the hope that readers with a strong interest in caves will become involved in cave conservation, for the caves we have today are a fragile resource for a rapidly growing population of cave and karst users.

I have deliberately tried to simplify some of the finer details of karst processes so that this book will appeal to the many with a keen, developing interest rather than to the few with detailed knowledge of small parts of the world of caves. To cater to the latter, I have provided some further reading in the specific scientific literature where these topics may be plumbed to greater depths. Where possible I have tried to stick to plain English terms but of necessity some terms from other languages have been used where these provide succinct words for complex concepts.

I am deeply grateful to Julie Kesby for her editorial skills. I am greatly indebted to Paul Ballard who has produced all of the figures for this book. I am also greatly in debt to Anne Cochrane and

Russell Drysdale for their considerable help in compiling and anno-
tating reference material. The work was also greatly assisted by a
special research grant from the University College, University of
New South Wales. Finally, I thank Caroline Richmond, my editor,
for her sound advice, tolerance and thoroughness.

I have been fortunate enough to visit many caves over the last
twenty-five years; among the goodly band of international
speleologists I must acknowledge my particular debt to Mike
Bourke, Derek Ford, Albert Goede, John Gunn, Ernst Holland,
Julia James, the late Joe Jennings, Kevin Kiernan, Stein-Erik
Lauritzen, Ian Millar, the late Jim Quinlan, Henry Shannon, Geary
Schindel, Dingle Smith, Andy Spate, Alan Warild, John Webb,
Tony White, Paul Williams, Steve Worthington and Yuan Daoxian.

The following people kindly provided photographs to supplement my own: Gareth Davies, Stefan Eberhard, John Gunn,
Andrew Lawrence, Franz-Dieter Miotke, Peter Serov, Andy Spate,
Tony White. Their individual contributions are acknowledged in
the plate captions.

I am also grateful to the following for permission to reproduce
copyright material:

Faber and Faber Ltd for permission to reproduce lines from *In
Praise of Limestone by W. H. Auden.*

Figures 1.2, 3.9, 3.11, 3.12, 3.14, 3.20, 3.22: from P. Courbon, C.
Chabert, P. Bosted and K. Lindsley, *Atlas of the Great Caves of the
World.* © Cave Books.

Figures 2.1 and 2.2: from P. L. Smart and S. L. Hobbs, Characteristics of carbonate aquifers: a conceptual basis. In *Proceedings of
the Environmental Problems in Karst Terranes and their Solutions.*
Copyright 1986. Reprinted by permission of National Ground
Water Association.

Figures 2.3, 2.4 and 2.12: from S. R. H. Worthington, Karst
Hydrogeology of the Canadian Rocky Mountains. Unpublished
PhD thesis, McMaster University. Courtesy S. R. H. Worthington.

Figure 2.6: from J. Jankowski and G. Jacobson, Hydrochemistry
of a grounderwater-seawater mixing zone, Nauru Island, Central
Pacific Ocean. *BMR Journal of Australian Geology and Geophysics*
12, 51–64. Courtesy Australian Geological Survey Organisation.

Figures 2.7 and 4.1: from *Geomorphology and Hydrology of Karst
Terrains* by W. B. White. Copyright © 1988 by Oxford University
Press, Inc. Reprinted by permission.

Figure 2.10: from *Journal of Hydrology* 61, P. W. Williams, The
role of the subcutaneous zone in karst hydrology, 45–67, © 1983.
With kind permission of Elsevier Science – NL, Sara
Burgerhartstraat 25, 1055 KV Amsterdam, The Netherlands.

Figures 2.16, 3.1, 3.16, 3.18, 3.19, 6.4: from D. C. Ford and P. W.
Williams, *Karst Geomorphology and Hydrology.* Copyright 1989.
Reprinted by permission of Chapman & Hall Ltd.

Figures 2.17, 2.18, 2.19: from J. F. Quinlan, R. O. Ewers and T. Aley, *Practical Karst Hydrology, with Emphasis on Ground-Water Monitoring. Field Excursion: Hydrogeology of the Mammoth Cave Region, with Emphasis on Problems of Ground-Water Contamination.* Reprinted by permission of National Ground Water Association. Copyright 1991. All rights reserved.

Figures 3.5 and 3.6: from J. Gunn, Solute processes and karst landforms. In S. T. Trudgill (ed.), *Solute Processes.* Copyright 1986. Reprinted by permission of John Wiley & Sons, Ltd.

Figure 3.8: from P. W. Williams and R. K. Dowling, Solution of marble in the karst of the Pikikiruna Range, northwest Nelson, New Zealand. *Earth Surface Processes and Landforms* 4, 15–36. Copyright 1979. Reprinted by permission of John Wiley & Sons, Ltd.

Figures 6.5 and 6.6: from M. Gascoyne, D. C. Ford and H. P. Schwarcz, Rates of cave and landform development in the Yorkshire Dales from speleothem age data. *Earth Surface Processes and Landforms* 8, 557–68. Copyright 1983. Reprinted by permission of John Wiley & Sons, Ltd.

Figure 7.7: from Y. Y. Shopov, D. C. Ford and H. P. Schwarcz, Luminescent microbanding in speleothems: high resolution chronology and paleoclimate. *Geology* 22, 107–10. Courtesy Geological Society of America.

Figure 8.1: from F. G. Howarth and F. D. Stone, *Pacific Science* 44 (2), 207–18, with permission of the publisher. © 1990 by University of Hawaii Press.

Figure 9.1: from P. W. Williams, Environmental change and human impact on karst terrains. In P. Williams (ed.), *Karst Terrains: Environmental Changes and Human Impact. Catena Supplement* 25, 1–20. Courtesy Catena Verlag.

Every effort has been made to trace the copyright holders of illustrations used in this book. However, if any have been overlooked the author and publishers will be pleased to make the necessary arrangements at the first opportunity.

ONE

The Cave System and Karst

What is Karst?

We can define the realm of caves in terms of a suite of landforms and associated geomorphic processes. Karstic terrain is commonly characterized by closed depressions, subterranean drainage and caves. This terrain is formed principally by the *solution* of the rock, most commonly limestone and its close relatives. But solution of rocks occurs in other lithologies, in particular in other carbonates such as dolomite, in evaporites such as gypsum and halite, in silicates such as sandstone and quartzites, and in some basalts and granites where conditions are favourable. All these are therefore true karst. Karstic terrain can also develop by other processes – weathering, hydraulic action, tectonic movements, melt water and the evacuation of molten rock (lava). Because the dominant process in these cases is not solution, we can choose to call this suite of landforms *pseudokarst*. This fundamental dichotomy is expanded in Table 1.1.

What is a Cave?

Your choice of a definition will depend on your dominant interest in these widespread and fascinating landscape features. A strictly scientific definition would be that *a cave is a natural cavity in a rock which acts as a conduit for water flow between input points, such as streamsinks, and output points, such as springs or seeps* (White 1984). It seems that once this type of conduit has a diameter larger than 5–15 mm, the basic form and hydraulics do not change much, though the diameter can be as much as 30 m. The range of minimum diameters allows turbulent flow, optimizing the solution of rock, and effective sediment transport. *Solution voids* that are not connected to inputs and outputs are types of isolated vugs; they may act as targets for developing cave systems. Small conduits less than 5 mm diameter but connected to an input or output or both

Table 1.1   Karstic terrains and processes

KARST	Rock type	Processes	Examples
	carbonates, e.g., limestone, dolomite	bicarbonate solution	Réseau Jean-Bernard, France; Mammoth Cave, USA
	evaporites, e.g., gypsum, halite, anhydrite	dissolution	Optimisticheskaja, Ukraine; Mearat Malham, Mt Sedom, Israel
	silicates, e.g., sandstone, basalt, granite, laterites	silicate solution	Sima Aonda and Sima Auyantepuy Noroeste, Venezuela; Mawenge Mwena, Zimbabwe
PSEUDOKARST			
	basalts	evacuation of molten rock	Kazumura Cave, Hawaii; Cueva del Viento, Islas Canarias; Leviathan Cave, Kenya
	ice	evacuation of melt water	glacier caves, e.g., Paradise Ice Caves, USA; Nilgardsbreen, Norway
	soil, especially duplex profiles	dissolution and granular disintegration	soil pipes, e.g., Yulirenji Cave, Arnhemland, Australia; Malabunga Caves, New Britain, Papua New Guinea
	most rocks, especially bedded and foliated rocks	hydraulic plucking, some exsudation	sea caves, e.g., Fingal's Cave, Scotland; Remarkable Cave, Tasmania
	most rocks	tectonic movements	fault fissures, e.g., Dan yr Ogof, Wales; Onesquethaw Cave, USA
	sandstones	granular disintegration and wind transport	rock shelters, e.g., Tassili, Algeria; Ubiri Rock, Kakadu, Australia
	many rocks, especially granular lithologies	granular disintegration aided by seepage moisture	tafoni, rock shelters and boulder caves, e.g., TSOD and Greenhorn Caves, USA

are called *protocaves*. These precursors of cave systems may carry seepage water or groundwater, and at the output may allow the formation of weathering hollows. These may coalesce to form rock shelters.

A simpler, non-scientific definition would be that *caves are natural cavities in a rock which are enterable by people*. This implies a minimum size of about 0.3 m diameter. These are the caves that we can explore, map and directly study (see plate 1.1). Most of the caves we study are in limestone and its related carbonate rocks, but significant caves (formed by a variety of processes) are also found in sandstones, evaporites such as gypsum, and basalt and granites. There are caves in the Antarctic ice and in the partially cemented dust of the loess plateau of China. In this book we will concentrate

Plate 1.1
A caver about to enter the strongly joint-guided passage of the Balmain Drain, in the Atea Kananda river cave of Papua New Guinea. Photo by Tony White.

on the caves that form in limestone, because they are the most numerous, certainly the largest, and we know most about them. The advent of cave diving has dramatically expanded the number and types of caves we can study, and information from these drowned conduits is causing some major revisions in thinking about caves. But most of the information in this book has been gained by the patient and often painful progress of cave scientists crawling, climbing and swimming in the subterranean world over the last century.

This book is based around three propositions.

First, that caves are a measure of the intensity and persistence of the karst process. As well as the continued efficiency of the limestone solution process through time, cave development is affected by tectonic activity and sea-level change. If the solution process has operated efficiently through time, then extensive caves will be found in a limestone massif. If that process has been severely interrupted by glaciation, aridity or sea-level change, then these effects will be reflected in cave morphology. Thus the extensive caves of the Nullarbor Plain, Australia, probably formed as large phreatic tubes under conditions of more effective rainfall in the late Tertiary. The progressive aridity of the Australian continent since that time has resulted in a change in the process regime, so that today the dominant process in the caves is collapse aided by salt wedging. Deepening of the caves has been enhanced by sea-level lowering in the Pleistocene, and today there are extensive flooded tunnels explored for up to 6 km. All these processes are reflected in cave morphology, and we have the prospect of dating these events by analysing the calcite, gypsum and halite formations found in the caves. In regions where climatic change has been minimal, such as the ever-wet tropics, then variation in cave development may reflect regional uplift patterns. In the karst towers of China there are caves at various levels – right up to the summits – which have been abandoned by their streams as the valleys have incised into rapidly rising plateaux. A more extreme example is provided by the alpine caves of the Canadian Rockies, which hang hundreds of metres above the valley floors and have been abandoned as the cordillera has risen over geologic time. Finally, the fluctuations in sea level during the Quaternary have produced cave development below present mean sea level, such as the Blue Holes of the Bahamas.

Secondly, that caves tend to integrate both surface and underground geomorphic processes. Solutes and sediments from the non-karst catchment of a cave combine with karstic solutes and sediments; these may be homogenized and lose their identity, or may be deposited in discrete units, such as flood deposits. This is somewhat akin to the way in which the products of catchment processes are integrated within a lake basin, and combine with processes in the water column and on the bottom to produce distinctive physical, chemical and biological properties (Oldfield 1977). Quaternary science can be applied to caves as well as lakes, and the caves provide us with an array of information on landscape processes on timescales ranging from yesterday to millions of years.

Thirdly, that once these products of surface and underground processes enter the cave system, they are likely to be preserved with minimal alteration for tens of millennia, perhaps even millions of years. In the near-constant temperature and humidity of the cave, weathering processes are reduced in intensity compared with the surface environment. The normal deepening of valleys leads to incision within the cave, leaving deposits in abandoned high-level passages. These are out of the reach of all but the largest floods. The unctuous clays common in caves have a great deal of resistance to erosion: once deposited, it is very difficult to erode cave sediments. Thus caves can be regarded as natural museums in which evidence of past climate, past geomorphic processes, past vegetation, past animals and past people will be found by those who are persistent and know how to read the pages of earth history displayed for them.

Where are the Deepest and Longest Caves?

Tables 1.2 and 1.3 were correct in December 1995, but the listing of the deepest caves may change radically due to the fast pace of cave exploration and mapping. Currently most of the world's deep caves are in Europe, especially in the younger mountain ranges of Western and Central Europe. Yet over two decades of cave exploration in the tropics is starting to yield very significant caves, especially in Sarawak and Mexico. The longest caves are dominantly in flat-lying limestones or gypsum and have been the scene of systematic exploration and mapping over many decades. It is unlikely that their ranking will change radically in the near future, but the prospect of its doing so is one of the most tantalizing facets of speleology.

What processes operate to form these major subterranean landforms? These karst landscapes are complex three-dimensional integrated natural systems comprised of rock, water, soil, vegetation and atmosphere elements. Nearly all of the karst solution process is moderated by factors operating on the surface of the karst and in the upper skin of the rock. Surface vegetation regulates the flow of water into the karst through interception, the control of litter and roots on soil infiltration, and the biogenic production of carbon dioxide in the root zone. The metabolic uptake of water by plants, especially trees, may regulate the quantity of water available

Table 1.2 The twenty deepest caves of the world (as at December 1995)

Name	Country	Depth
1 Réseau Jean-Bernard	France	−1602 m
2 Lamprechtsofen-Vogelschacht	Austria	−1535 m
3 Gouffre Mirolda/Lucien Bouclier	France	−1520 m
4 Shakta Vjacheslava Pantjukina	Georgia	−1508 m
5 Sistema Huautla	Mexico	−1475 m
6 Sistema de la Trave	Spain	−1441 m
7 Boj-Bulok	Uzbekistan	−1415 m
8 Illaminako Ateenoko Leizea (BU56)	Spain	−1408 m
9 Lukina Jama	Croatia	−1393 m
10 Sistema Cheve	Mexico	−1386 m
11 Snieznaja-Mezennogo	Georgia	−1370 m
Ceki 2 (Cehi)	Slovenia	−1370 m
13 Réseau de La Pierre Saint-Martin	France/Spain	−1342 m
14 Sieben Hengste-hohgant Höhlensystem	Switzerland	−1324 m
15 Gouffre Berger-la Fromagère	France	−1278 m
16 Cosanostraloch-Berger-Platteneck Höhle	Austria	−1265 m
17 Torca dos los Rebecos	Spain	−1255 m
18 Pozo del Madejuno	Spain	−1254 m
19 Abisso Paolo Roversi	Italy	−1249 m
20 Vladimira Iljukhina System	Georgia	−1240 m

Table 1.3 The twenty longest caves of the world (as at December 1995)

Name	Country	Length
1 Mammoth Cave system	USA	531 069 m
2 Optimisticheskays	Ukraine	183 000 m
3 Holloch	Switzerland	156 000 m
4 Jewel Cave	USA	144 800 m
5 Sieben Hengste-hohgant Höhlensystem	Switzerland	126 000 m
6 Wind Cave	USA	113 300 m
7 Ozernaja	Ukraine	111 000 m
8 Lechuguilla Cave	USA	106 000 m
9 Fisher Ridge Cave	USA	104 000 m
10 Gua Air Jernih (Clearwater Cave)	Malaysia	101 500 m
11 Ojo Guarena	Spain	97 400 m
12 Coume d'Hyouernède	France	89 496 m
13 Zolushka	Moldavia	85 500 m
14 Sistema Purificacion	Mexico	79 100 m
15 Ease Gill Cave system	Great Britain	70 500 m
16 Raucherkarhöhle	Austria	70 000 m
Hirlatzhöhle	Austria	70 000 m
18 Friar's Hole Cave	USA	69 234 m
19 Organ Cave	USA	60 510 m
20 Red del Silencio	Spain	60 000 m

to feed cave formations. Water is the primary mechanism for the transferral of surface actions on karst to become subsurface environmental impacts in the caves.

Pre-eminent among these karst processes is the cascade of carbon dioxide from low levels in the external atmosphere through greatly enhanced levels in the soil atmosphere to reduced levels in cave passages. Elevated soil carbon dioxide levels depend on plant root respiration, microbial activity and a healthy soil invertebrate fauna. This cascade must be maintained for the effective operation of karst solution processes, and is clearly dependent on biological processes. Thus we cannot consider caves and karst as geomorphic systems in isolation from biological processes. In this book we examine the processes, development and management of caves from both geomorphic and biological viewpoints. The main focus will be on the geomorphology of caves, so in common with surface landforms we must consider the flows of mass and energy through subterranean realms.

Caves as Geomorphic Systems

A variety of materials enter caves, travels through them or are stored temporarily or permanently, and finally leave the cave system at springs. The range of materials is illustrated in figure 1.1. These can be conveniently divided into a number of inputs and outputs, though in reality the boundaries between these conceptual compartments are somewhat blurred. In addition, not all of the

Figure 1.1
A hypothetical cave system showing the potential types of inputs and outputs from the system. Each of these may be discontinuous in space and time. Quantification of some elements of the cave system, especially those below the water table, may be difficult if not impossible.

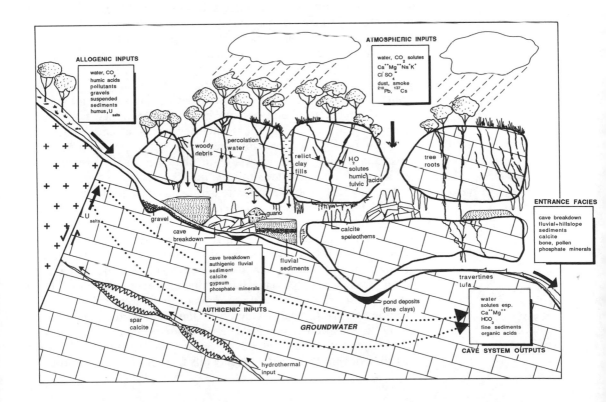

materials will be present in any one cave system, nor will a particular material be present for all of the cave's history. These absences raise interesting questions for geomorphologists about changes in the process regime in an area, and allow inference of climatic change, tectonic activity or, latterly, human impact.

A good deal of the material to be seen in any cave is derived from its catchment, transported into the cave by fluvial, mass movement, aeolian or glacial processes. This transport may be episodic, and material may have been temporarily stored in small terraces or floodplains prior to entering the cave. Thus the material may have already been altered by weathering or sorting since leaving its source area. The typical materials which derive from the catchment are water, of course, dissolved gases such as carbon dioxide, fluvial gravels and fine sediments (sands, silts and clays), humus and dissolved organic acids, and a range of solutes which will depend on the catchment lithology. Perhaps the most important solutes for the student of karst are calcium and magnesium cations, the ephemeral bicarbonate anions, the uranium series anions, and an increasing range of pollutants including herbicides, sewage, petroleum products and some heavy metals. Others may help to characterize the karst water in tracing experiments, but are of secondary importance in understanding processes.

A range of materials enter the cave as *atmospheric inputs*: these are transported as solutes in rainwater, as hydrometeors, or as dust. Again water and carbon dioxide are important, along with cations and anions gained from sea spray or from cyclic salts. Radioactive fallout containing the important isotopes lead-210 and caesium-137 may also enter the cave directly from the atmosphere, and become incorporated in cave sediments or biota. Dust and smoke particles are also important materials; at Yarrangobilly in New South Wales, bushfire smoke is often seen being sucked into caves, and thin black smoke layers can be seen in sectioned flowstones. Elsewhere in Australia, the Kimberley caves often contain red quartz sand layers, blown off the Great Sandy Desert in Western Australia during more arid climate phases than the present. Together all these form the *allogenic inputs* to the karst system.

A number of processes operate within any cave to produce debris and chemical precipitates. These *authigenic inputs* are spatially and temporally quite variable. The most common is cave breakdown, the angular debris that may range in size from pebbles to house-sized blocks. Breakdown is heavily influenced by the mechanical strength of the limestone and the geological structure. It tends to occur after a cave passage has drained, because the roof of the passage loses the hydrostatic support of the water. The stoping of the passage produces a characteristic flat arch, with a pile of debris below it. This debris pile may be later capped by calcite flowstone or stalagmites. The subsequent erosion of breakdown by cave streams, or the slow release of clay fills from roof crevices, provides authigenic fluvial sediments. Finally, although calcite dominates the chemical precipitates in most caves, a wide range of minerals may be deposited, including gypsum, halite, and a complex array of phosphatic minerals associated with bat or bird guano.

Cave waters emerge at springs or seepages. At these points, the *cave system outputs* are water, perhaps some dissolved gases (carbon dioxide and/or methane), the products of limestone solution, some fine clay sediments, and organic acids leached from guano. The solutes may be augmented substantially from solution processes operating below the water table in the phreatic zone, and carried to the outlet spring in the steady flow of groundwater.

A curious mixture of allogenic and authigenic materials becomes preserved as *entrance facies* deposits in the mouths of caves and in rockshelters. These sites are the targets of intensive excavation and research by palaeontologists and archaeologists. As well as cave breakdown, whose formation is enhanced by temperature and humidity fluctuations in the entrance zone, relict fluvial and hillslope sediments are well preserved there. These sediments may become coated with calcite deposits, enhancing their preservation and providing a dateable material. The bones of animals, including people, become incorporated into the sediments, as well as plant remains such as pollen and, rarely, leaves, nuts and seeds. The leaching of bones may provide opportunities for the formation of phosphatic minerals. Near to valley sides, tree roots may penetrate the limestone to great depths along fissures, and act to channel water and humic substances deep into the cave.

Although the flux of solutes through cave systems has been investigated worldwide, there have been few attempts to estimate the flux of all matter through a cave system. This is due partially to the extreme difficulty of quantifying some items – cave breakdown, for example – and of estimating the age of many or all of the components. One set of estimates has been prepared by Steve Worthington for the Friar's Hole system of West Virginia, USA (see figure 1.2 and table 1.4). This is an extensive system which is not well endowed with formations and reflects episodes of cold climate conditions throughout its long history of some 4 million years. Limestone accounts for only 3 per cent of its surface catchment, the remainder of which is underlain by shale, sandstone and clay-rich limestone. These weather to produce abundant regolith which can be transported. Not surprisingly, 95 to 98 per cent of the total mass flux through the cave has been this allogenic fluvial sediment.

This result is probably representative of mid- to high-latitude caves where speleothem deposition has been restricted during glacials and may decline seasonally today. Older caves, long abandoned by their streams, might yield higher proportions of calcite speleothems. Caves in the tropics would probably yield higher proportions of organic matter, both from allogenic suspended sediment and authigenic bat guano and its byproducts. The proportion

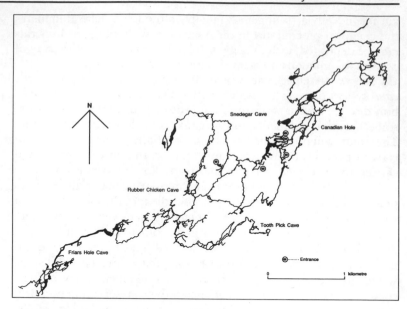

Figure 1.2
*Plan of the Friar's Hole
system, West Virginia,
USA. Several caves have
been linked by exploration
since 1960; there are ten
entrances presently known.
After Courbon et al.
(1989, 97).*

Table 1.4   Friar's Hole Cave system, West Virginia (from Worthington 1984)

Length 68 km, depth −188 m

Total volume of known cave				2700 × 10³ m³
	− open cave			1800 × 10³ m³
	− debris filled			900 × 10³ m³
Catchment area				85.7 km²
	− includes limestone area			2.6 km²

Mass flux	Total 10³ m³	%	Present 10³ m³	%
Dissolved limestone:				
1 from catchment	57 300	1.9	trace	−
2 from cave	2400	0.0008	trace	−
Cave breakdown	1000	0.0004	280	31
Allogenic fluvial sediment	3 000 000	98.0	600	67
Authigenic fluvial sediment	400	0.0001	20	2
Organic matter	100	0.00003	0.001	negligible
Calcite speleothems	<1	negligible	0.15	negligible
Gypsum speleothems	0.001	negligible	0.001	negligible
Aeolian deposits	<1	negligible	0.15	negligible

of open cave to debris-filled cave may vary significantly: deep guano
fills, many up to 20 m thick, occupy most of the cavity in Sarawak
and Australian caves, and the open space through which we walk
is but a small component. As a consequence, cave histories
gained solely from examination of evidence in the open, air-filled
cave may neglect crucial information contained in or buried by the
sediments.

Now the Details . . .

This book is organized in three sections. In this first section, contemporary processes of cave formation are examined. Caves can be regarded as three-dimensional networks in which individual links may become abandoned or may develop through time. The nature of a cave at any time is a function of inheritance of previous states and the contemporary climatic and tectonic setting. The state of the hydrological network may be fully phreatic, fully vadose or a combination, depending on the position of the water table in a cave. The nature of individual linkages (passages) may change depending on the frequency of floods and their exploitation of existing or potential conduits in the rock. In caves with a temporary or permanent air space, the nature of the cave climate will determine the kind of speleothem deposits that form (calcite, ice, gypsum, halite). There is also a feedback between climate and the karst solution process.

The roles of the kinetics of limestone solution, of rock architecture and cave collapse in the evolution of the position of the subterranean water table and in cavern enlargement are outlined. The effects of brine and gypsum on cave formation are regionally important. The formation of weathering caves and pseudokarst caves is another aspect of the subject. The final form of a cave owes much to both the purity of the limestone and the network of fissures which dissect the rock. Following a brief treatment of limestone lithology and structural variation, examples of cave passage shapes in different limestone settings are given. Geological history plays a key role and may override other factors. Some of the largest and deepest caves are formed by combinations of these factors.

The second section of the book deals with past processes and their products. The diversity of cave formations has long fascinated people. In Chapter 4 the basic mechanisms of calcite deposition are given, including the roles of trace elements and cave biota in providing shape and colour variation. Effects of tectonics and hydrologic change on cave formations are outlined. Gypsum and halite speleothems are briefly reviewed. In the tropics biogenic deposits such as guano are dominant and have economic significance.

As we have seen, cave sediments account for the bulk of mass flux through a cave. Caves can be seen as underground gorges and floodplains, in which sedimentation proceeds in modes analogous to those of surface fluvial systems. Cave sediments may be of external or internal origin; various types and properties will be described (glacial, fluvial, aeolian, biogenic). Relationships exist between cave sediment structures and depositional energy, and interactions between cave sedimentation and hydrology are relevant. Reconstruction of cave hydrology from sediment surface textures using scanning electron microscopy is an important recent development.

In the last two decades a bewildering array of dating techniques have become available to the cave scientist. The most important of these is uranium series dating. This technique will be outlined in detail, with examples from Europe and North America. Other techniques such as electron spin resonance (ESR) and thermoluminescence (TL) will be outlined briefly, while the use of palaeomagnetism and radiocarbon will be considered. These dating techniques have radically altered scientific thought about caves and landscape evolution. Calcite speleothems can be used as palaeo-thermometers through the technique of oxygen isotope analysis. Carbon isotope analysis on speleothems has potential for determining changes in the surface vegetation above the cave. This exciting topic is the fastest developing field in karst research.

Finally, cave ecosystems and their management are considered in the third section of the book. Physiological and evolutionary adaptations to cave living occur in many phyla of the animal kingdom, and also in both flowering plants and fungi. These adaptations are broadly classified through the scheme of cave life – trogloxene, troglophile, troglobite. The basic characteristics of the cave ecosystem and its constituent trophic levels are described in Chapter 8. This relates back to the basic concepts of mass and energy flow in the karst system. The role of external energy sources is critical for cave life. Cave biota are dependent on periodic inputs of nutrients, usually swept in by floods. They are also severely disadvantaged by quite minor disturbances. Thus they have low resilience in the face of a change to the cave ecosystem.

Caves have long been of importance to people – for shelter, water supply and food and as objects of veneration. Recent uses include mining of cave formations and guano (American Civil War), hydro-electricity and as sanatoria. Increasing cave tourism presents problems owing to the irreversible degradation of cave ecosystems and alteration of cave microclimates. The addition of energy sources (heat, lint and dead skin cells) alters the trophic status of caverns. Caves must be classified in terms of their limits to acceptable change for the proposed use (recreation, resource extraction). This classification must be dynamic to allow for seasonal changes in cave function and use. Cave lighting and pathways must be designed to minimize the effects on cave microclimates and biota. Subtlety and effective interpretation are the key tools of cave managers. Only through public education can the aims of ecologically based cave management be achieved. There is now a global network of cave scientists considering these problems. These themes, both above and below ground, are addressed in chapters 9 and 10.

References

Courbon, P., Chabert, C., Bosted, P. and Lindsley, K. 1989: *Atlas of the Great Caves of the World*. St Louis: Cave Books, 368 pp.
Oldfield, F. 1977: Lakes and their drainage basins as units of sediment

based landscape reconstruction. *Progress in Physical Geography* 1, 460–504.

White, W. B. 1984: Rate processes: chemical kinetics and karst landform development. In R. G. La Fleur (ed.) *Groundwater as a Geomorphic Agent*. Boston: Allen & Unwin, 227–48.

Worthington, S. R. H. 1984: The Palaeodrainage of an Appalachian Fluviokarst: Friar's Hole, West Virginia. MSc. thesis, McMaster University.

TWO

Cave Hydrology

Basic Concepts in Karst Drainage Systems

In normal landscapes small streams unite to form larger rivers, which take water, solutes and sediments to the sea in an obviously integrated drainage network. In karst landscapes this surface integration is disrupted by the formation of small, centripetal drainage basins, with the landscape punctuated partially or wholly by closed depressions. Below the surface a complex network of fissures and conduits carries the water and erosion products to springs where this karst drainage network unites with the surface again.

The karst drainage network thus includes surface elements such as vegetation, soil, regolith, and closed depressions which regulate the quantity and quality of water passing underground. It encompasses the subterranean elements of various sizes and shapes of pores, fissures and conduits (or caves) which transmit and store water and act as repositories for chemical precipitates, organic and inorganic sediments. Finally, its output is regulated by springs at the karst margin, which may be enterable caves, flooded tubes or diffuse seepages in a stream bed. The most important attributes of the karst drainage system are the intimate connections between surface and underground drainage, and the rapidity with which surface water may enter and percolate down through the karst rock. This is related to the greater fissure density in karstic rocks, which enhances their ability to transmit and store water.

Karst drainage systems tend to be steeper than their surface counterparts because the underground diversion of water short-circuits much of the rock mass. In non-karst landscapes the retreat of a river profile upstream extends from a base level (sea, lake or resistant rock barrier) with substantial lateral and vertical erosion. The long profile of the stream is determined by the rate of downcutting relative to the rate of uplift. In karst the same is essentially true, but underground water tends to reach base level quicker, producing a steeper gradient and relatively dry rock masses above the caves. The lowest caves in a vertical sequence are more susceptible to modification owing to local changes of base level. The degree and interconnectedness of cracks and fissures in the rock have a great influence on the gradient of the underground stream. The nature of this rock porosity is considered next.

Karst Aquifers

All rocks are capable of transmitting water to some degree through the pore spaces between the mineral grains. Rock formations capable of storing large amounts of water are called *aquifers*, a term widely used in hydrology and economic geology. Aquifers are characterized by their thickness, area, water storing or transmitting properties and the mean water quality. In this text the term karst drainage system is preferred because it integrates the entire karst system including components such as soil, regolith, recharge and discharge, which generally fall outside those considered in aquifer behaviour. It also avoids any assessment of either the quantity or quality of the water involved. In the scientific literature the terms are often used synonymously.

Water infiltrates into the karst through the soil and regolith, moving down under gravity until it reaches the zone in which all the pores are filled with water. The surface separating air-filled from water-filled pores is called the piezometric surface or *water table*, and its contours can be mapped from elevations in boreholes. The saturated rock below the water table is called the *phreatic* zone, while the largely air-filled rock above the water table is called the *vadose* zone. If the vadose zone connects directly with the surface, so that the water table can respond freely to rainfall and spatially varying infiltration, then the aquifer is said to be *unconfined*. This is common in karstic rocks where there is not an overlying rock layer of lower permeability. If there is an overlying, impermeable rock layer then the water may be in a *confined aquifer* and may be under some pressure. The water table cannot respond to varying infiltration, and instead there is a potentiometric surface which corresponds to the varying pressure head developed under the confining rock formation. Any well or bore drilled into the rock will have a water level at this potentiometric surface. If this lies above the ground surface, water will flow freely from the well and we have an *artesian aquifer*.

Within a karst water will move slowly from recharge areas, where the water table is highest, to discharge areas, where the water table is lowest. This movement occurs along a *hydraulic gradient* which is the height difference (or head) divided by the distance. Clearly the rate of water movement will be limited by this factor and by the porosity and permeability of the karst rock.

Porosity and Permeability of Karstic Rocks

The porosity of a rock is the volume fraction occupied by voids or pores. It is dimensionless and can be defined as

$$\theta = \frac{Vpores}{V} = \frac{1 - Vmin}{V} \qquad (2.1)$$

where V is the bulk volume of the rock and Vmin is the net volume of mineral grains.

Permeability is the ability of a rock to transmit fluid, usually water. A rock must be porous to transmit fluids, but the pores must be connected for this to occur. In a uniform porous medium the fluid flow is governed by Darcy's law, which states that the flow through a porous medium depends on the hydraulic gradient:

$$q = \frac{Q}{A} = -K\frac{dh}{dl} \qquad (2.2)$$

where q is the unit flow in $m^3 s^{-1}$ across an area, A (m^2), of the medium. The hydraulic gradient, dh/dl, is dimensionless, so both K and q have units of velocity, $m s^{-1}$. The hydraulic conductivity, K, depends on both fluid and rock properties. Darcy's law is usually stated as:

$$q = \frac{-Nd^2\rho g}{\eta}\frac{dh}{dl} \qquad (2.3)$$

where g = gravitational acceleration, ρ = fluid density, d = grain diameter, η = fluid viscosity, and N is a dimensionless shape factor.

The quantity Nd^2 is the effective permeability of the rock; this is difficult to calculate and so is usually based on laboratory measurements or on aquifer characteristics. It is usually expressed in area units as cm^2 or m^2. For limestones the range of values is from 10^{-5} to $10^2 m^2$ or more, while for sandstones it may range from 10^{-8} to $10^{-3} m^2$ (Ford and Williams 1989, 153).

Primary porosity is the result of the packing of mineral grains during the formation of a rock. This may be modified by solution or deposition during diagenesis. Thus primary porosity is rarely uniform in a rock. Through time the formation of fractures, enlarged joints or bedding plane partings produce connected openings through which water can circulate. Brittle but strong rocks such as limestone, dolomite, sandstone, and granite readily crack and these fractures tend to remain. Weaker rocks such as shales, mudstones and chalk may crack but the fractures often seal owing to the pressure of overlying strata. Together the various fractures, joints and bedding plane partings constitute *secondary or fissure porosity*. This is highly variable in the rock mass and allows for the concentration of water flow in areas of higher permeability. Through time some of these fissures may thus enlarge in a karst to form conduits – which is sometimes termed the *conduit porosity*. These conduits may range in size from small solutionally widened fractures or fissures, 1 to 100 mm wide, to major trunk passages 30 m in diameter (see table 2.1).

Table 2.1   Porosity types and karst aquifer properties

	Primary porosity	Secondary porosity	Conduit porosity
Components	Intergranular pores Vughs Mineral veins	Linked joints and fractures Bedding plane partings Connected mineral veins	Open channels and pipes of variable size and shape
Homogeneity	Usually isotropic	Usually anisotropic due to fracture origins, often oriented	Highly anisotropic forming networks
Flow regime	Laminar	Laminar to just turbulent	Turbulent
Governing hydraulic law	Darcy	Hagen–Poseuille	Darcy–Weisbach
Water table	Well defined	Irregular surface	Often perched and at varying levels
Flow response to input water	Slow	Moderate	Rapid

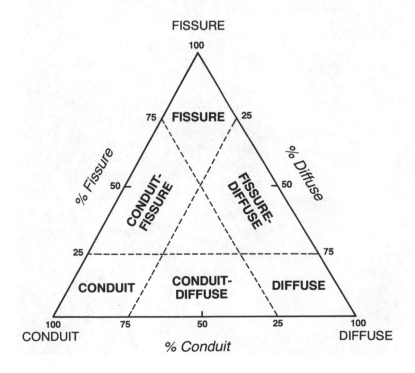

FISSURE

FISSURE

CONDUIT-FISSURE

FISSURE-DIFFUSE

% Fissure

% Diffuse

CONDUIT

CONDUIT-DIFFUSE

DIFFUSE

CONDUIT

DIFFUSE

% Conduit

Figure 2.1
Conceptual types of karst aquifers and their mixtures. From Smart and Hobbs (1986).

In reality any karst aquifer may be regarded as having elements of all three porosity types – diffuse, fissure and conduit (see figure 2.1). It is unwise to analyse any single karst hydrologic system assuming that one type prevails in water movement. Analysis of a karst aquifer assuming diffuse flow – an equivalent porous medium obeying Darcian flow – will generally underestimate flow velocities from recharge zone to spring by two or three orders of magnitude (Worthington 1991). This may have serious repercussions if pollutant movement is being predicted.

The presence of all three flow types may be seen in many active stream passages. The water flowing in the cave stream is typical of low stage conduit flow (see plate 2.3). Water dripping from straws and stalactites in the roof, perhaps aligned along a fracture or joint, is obeying fissure flow. Water seeping through the intergranular pores and emerging as wet patches on the roof is governed by diffuse flow. The magnitude and relative importance of each type in an individual karst hydrologic system will depend on the lithology and secondary porosity of the host rock. Thus in a porous rock such as chalk, diffuse flow may dominate, but there may be a significant amount of fracture flow. But chalk is a mechanically weak rock and conduits may not survive. In contrast marble has low primary porosity and thus diffuse flow may be negligible.

Plate 2.1
The streamsink of the Atea River enters Atea Kananda cave, Muller Range, Papua New Guinea. The waterfalls are perched on insoluble beds in the limestone.

Water will tend to move quickly in fractures and conduits from recharge zones to the springs.

Zonation of the Karst Drainage System

The karst drainage system may be usefully divided into a number of zones, each of which has distinctive hydraulic, chemical and hydrological characteristics. These zones are not mutually exclusive, and under radically altered flow conditions one zone may temporarily gain the properties of the one below it in the sequence. This sequence of zones is as follows:

Karst drainage system:

- Epikarst:
  - Cutaneous zone (surface and soil)
  - Subcutaneous zone (regolith and enlarged fissures)
- Endokarst:
  - Vadose zone (water unsaturated)
  - Phreatic zone (water saturated)

The upper portion of the vadose zone, where discrete threads of water in the subcutaneous zone join to form percolation streams, is often termed the percolation zone. In recognition of the transient boundaries between some zones, the lowest portion of the vadose zone within the range of floodwater-level fluctuations is termed the epiphreatic zone (or temporary phreas). It is intermittently saturated. The lower phreatic zone, at depths of a few hundred metres, is termed the bathyphreatic zone. The lowest portion of the phreatic zone, where water moves slowly in enlarging voids of varying size, is termed the nothephreas or stagnant phreatic zone.

Caves may be regarded as three-dimensional networks of conduits of varying size, from a few millimetres to tens of metres in diameter, which extend from inputs to outputs and transcend these zone boundaries, either at a single time or over the lifetime of a particular network. In this sense caves may be regarded as four-dimensional networks, in that there is a dynamic aspect to the network which may alter its structure and function during a single flood event or over geological time. The long-term evolution of cave networks is considered in more detail in the next chapter.

Defining the Catchment of a Cave

The karst drainage system resembles that of a surface stream or river in that it is comprised of a network of channels great and small which transport water, solutes and sediments from a single input or an array of inputs, often known as streamsinks, to single or multiple outputs, often known as springs. The surface analogues of these are the first-order streams in a drainage basin, where integrated flow commences, and the terminal basin of the catchment, where the channel debouches into a lake, an estuary or the sea. Whereas the catchment of a surface stream may be readily defined using aerial photography or satellite imagery, the catchment of a karst drainage system may be very difficult to define unequivocally.

Although some of the inputs to the drainage network may be discrete streamsinks whose connections can be proven by dye-tracing experiments, many will be diffuse feeders in the subcutaneous zone whose flow paths will vary depending on individual storm intensity and antecedent rainfall. These diffuse sources are fed by soil water infiltration and percolation on the karst surface, and by soil water throughflow from adjacent non-karstic rocks and regolith. Although some short distance tracing of these inputs has been carried out, elucidation of the uppermost elements of a drainage network from diffuse sources is extremely difficult. There is also the possibility that high-level, abandoned conduits or palaeokarstic features may be reactivated under exceptional rainfall conditions.

There is therefore some truth in the statement that the catchment of a karst drainage system is usually much larger than just the area of limestone outcrop and the obvious non-karstic contributing catchment. Breaching of surface drainage divides by karst drainage systems is very common, as the network may have formed under an older surface topography which bears little resemblance to that visible today, or the evolving karst drainage network may have captured adjacent networks. The details of these connections can be worked out only by careful dye tracing and by mapping of cave passages.

The catchment boundary is not a single line that can be represented on a map, but a zone which has a dynamic outer boundary dependent on local details of surficial geology and weather conditions. It is more useful to think of a core catchment area, within which flow will usually be directed to a particular cave network, and a peripheral or buffer catchment area which may be activated periodically. If the precautionary principle applies in karst research or management, then the larger catchment may provide a truer representation of the sources for the karst drainage network.

Hydraulics of Groundwater Flow in Karst

Early work on karst drainage systems emphasized two modes of flow: diffuse flow and concentrated or conduit flow. Atkinson (1977) demonstrated that both existed in the Mendip Hills, England, and that conduit storage was only about 3.5 per cent of the total baseflow storage in the karst drainage system. Conduits were, however, very important in draining the diffuse part of the karst system. From an analysis of spring water chemistry, Bakalowicz and Mangin (1980) concluded that this bimodal scheme was too simple to explain the observed variation, and from an evaluation of a much larger sample of karsts Atkinson and Smart (1979) suggested that an intermediate-sized void, the karst fissure, was important in some drainage systems. This led Atkinson (1985) to propose a three-member spectrum of water flow which has found general acceptance (see figure 2.1). Depending on the nature of the karst host rock and its geomorphic history, any karst drainage system may be placed in this scheme.

Smart and Hobbs (1986) elaborated this approach by suggesting that the properties of recharge, transmission and storage might help to position individual karst drainage systems in the spectrum (see figure 2.2). For recharge the end members are concentrated and diffuse inputs; for transmission, conduit and diffuse flow; and for storage, unsaturated and permanently saturated stores. Later consideration by Ford and Williams (1989) and by Worthington (1991) led to an appreciation of the importance of the role of spring aggradation in controlling drainage system behaviour, and of the role of fissure flow. The consequences of these three factors of

Figure 2.2
Conceptual scheme of sensitivity of karst aquifers to disturbance, for storage, recharge and flow types. Boundaries between fields 1 to 5 are approximate and intuitive. A crude measure of the diagonal from field 4 to field 1 is given by increasing values of the coefficient of variation (s.d./mean) of specific conductivity of the water. Modified from Smart and Hobbs (1986).

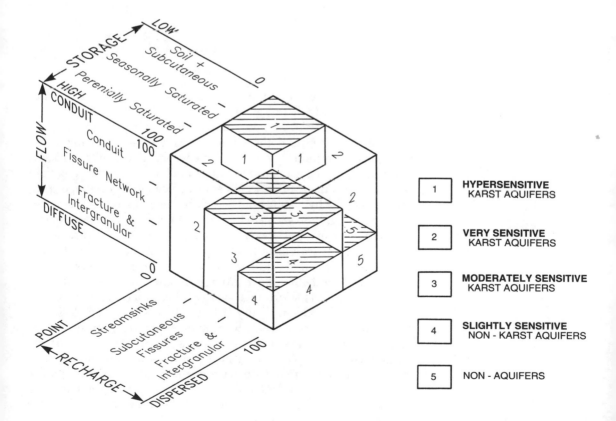

1	**HYPERSENSITIVE** KARST AQUIFERS
2	**VERY SENSITIVE** KARST AQUIFERS
3	**MODERATELY SENSITIVE** KARST AQUIFERS
4	**SLIGHTLY SENSITIVE** NON - KARST AQUIFERS
5	NON - AQUIFERS

recharge, flow and storage for spring behaviour will be considered later. It is first necessary to define the three modes of water flow in the karst drainage system.

Diffuse flow

In a homogeneous porous rock the flow of water can be characterized by Darcy's law, which has already been stated in equation 2.2.

The movement of water in a homogeneous aquifer can be regarded as being at right angles to the contours of the water table or piezometric surface, and this surface of hydraulic equipotential lines can be mapped from borehole or well level data. There is thus a flow net made up of streamlines which extend down the maximum slope or hydraulic gradient of the piezometric surface from the inputs at the upper boundary of the catchment to the springs or outlets at the lowest point. This flow net may descend to considerable depths below the spring level in any karst drainage system. If flow potential increases with depth then at some point the flow will converge and turn up to an area of lower potential, usually below the spring or surface river. A large conduit in the saturated zone will also have lower potential, and will thus serve as a target for the flow net. The rate of groundwater inflow per unit length along such a conduit can be estimated by

$$Q_i = \frac{2\pi K h_z}{2.3\log(2h_z/r)} \tag{2.4}$$

where h_z is the depth of the conduit of radius r below the water table (Freeze and Cherry 1979).

For Darcy's law to be applicable to karst systems, laminar flow must be occurring and there must be conservation of mass (the principle of continuity), where water entering a unit cube of rock is balanced by water leaving it. At higher velocities beyond a certain threshold, turbulent flow occurs and the velocity is no longer proportional to the hydraulic gradient. Turbulence also occurs in water flowing through porous media when the interconnected pore spaces are larger than fine gravels. Despite these problems, many groundwater hydrologists have applied the principles of Darcian flow to karst aquifers, and it remains the most commonly used method for modelling flow from borehole and well data. For Palaeozoic aquifers Worthington (1991) has noted that the velocity predicted from Darcian theory (10^{-7} to $10^{-5}\,\mathrm{m\,s^{-1}}$) is much less than that actually observed from dye-tracing experiments from well to well (average $0.02\,\mathrm{m\,s^{-1}}$). In part this is due to the fact that the conduits are the rapid flow elements in the karst drainage system and are fed by the much slower movement of water through the rock mass. Thus to understand the movement of water in karst it is necessary to look at all components of the underground drainage network.

Fissure flow

Where laminar flow occurs in a jointed or fractured rock, the hydraulic conductivity K (in m.s^{-1}) of a fissure with parallel sides can be determined from the equation

$$K = \frac{gw^3}{12v} \qquad (2.5)$$

where w is the width of the fissure and v is the kinematic viscosity of the liquid (Ford and Williams 1989, 132). The kinematic viscosity is heavily dependent on temperature, while it is apparent that a slight increase in fissure width will cause a dramatic increase in hydraulic conductivity. Thus laminar flow in an enlarging karst fissure may quickly become turbulent, and conduit flow conditions obtain.

Scaling up from a single fissure to a fractured aquifer is very difficult. Irregularities and variations in fissure width along the flow path make it difficult to arrive at a single representative width, and the dependence of flow on the cube of this value makes it a sensitive parameter (White 1988, 171). Finally, the geometry of complex fracture sets is often unknown and is very difficult to model in three dimensions. Worthington (1991) has established a relationship between flow path length and joint orientation (see figure 2.3).

Conduit flow (turbulent and laminar cases)

Active conduits are major determinants in karst drainage system behaviour and function. They are the foci for the discharge of tributaries in the karst, and they are the main vectors for groundwater flow through and discharge from the karst system. Unfortunately the flow of water in active conduits is very hard to observe directly (see plate 2.2) and even harder to model satisfactorily. There have been significant recent advances in our understanding of the structure and function of conduits, due largely to the efforts of cave divers and by carefully designed dye-tracing experiments. Inferences can be drawn from observations of drained conduits or caves, but Lauritzen et al. (1985) have warned that the study of fossil conduits 'is a direct parallel to the case of medieval anatomists dissecting a dead body rather than studying the physiology of the living organism.' But key questions such as the magnitude of storage and the velocity of water movement remain unresolved.

The flow of water through small tubes was studied by Hagen (1839) and later by Poseuille (1846). They discovered that the water flow or specific discharge μ was proportional to the hydraulic head loss by friction Δh along the tube.

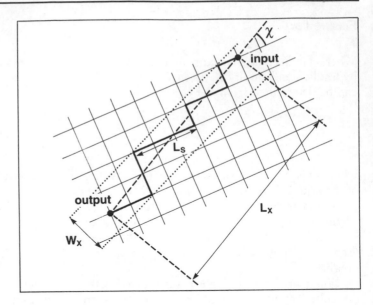

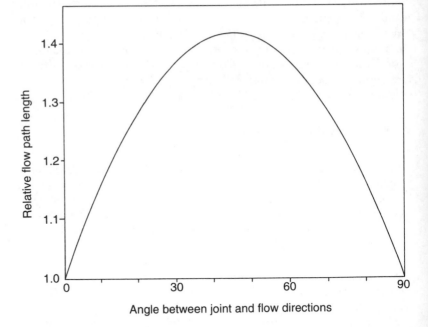

Figure 2.3
Relationship between flow path length and angle between joint and flow directions. L_x and W_x indicate the length and width of the flow zone respectively. From Worthington (1991).

$$u = \frac{\pi r^4}{8 l \mu} \cdot \Delta h \qquad (2.6)$$

where r is the radius of the tube, μ is the kinematic viscosity of the water, and l is the length of the tube. The hydraulic head loss Δh is the difference in head between one end of the tube and the other, where the head at a point is the sum of the pressure head, the depth

Plate 2.2
*Looking out of the
entrance of the horizontal
river cave Yo Chib,
Chiapas, Mexico. Photo by
Gareth Davies.*

of the point below the water table, and the elevation head, the
height of the point above the datum or base level.

Under laminar flow condition in small tubes, the discharge per
unit length can be calculated using the Hagen–Poseuille equation
(Vennard and Street 1976):

$$Q = \frac{\pi d^4 \rho g}{128 \mu} \cdot \frac{dh}{dl} \qquad (2.7)$$

where *dh/dl* is the head loss over a unit length and *d* is the diameter
of the tube.

From this, large tubes are much more conductive than small
ones. A tube 1 mm in diameter will conduct the same flow as 10 000
capillaries 0.1 mm in diameter. This means that a single enlarging

tube will rapidly capture the flow from smaller ones, and is of great importance for an evolving cave network (see plate 2.3).

The Reynold's number R_e is used to help identify the critical velocity at which the transition from laminar to turbulent flow takes place and beyond which Darcy's law is inapplicable. This may occur when R_e is greater than values of 1 to 10, but full turbulent flow is attained at high velocities and R_e values in the range 10^2 to 10^3. Reynold's number is expressed as

Plate 2.3
A phreatic tube with incised canyon, Flabbergasm Passage, Dan-yr-ogof, South Wales. This is a triple porosity passage – conduit flow in the tube, fissure flow feeding stalactites, and calcite deposition from diffuse seepage. Photo by Gareth Davies.

$$R_e = \frac{\rho v d}{\mu} \tag{2.8}$$

where v is the mean velocity of the fluid flowing through a pipe of diameter d.

When the flow in an enlarging tube becomes turbulent, a threshold is passed beyond which the Hagen–Poseuille equation becomes invalid and discharge is better estimated using the Darcy–Weisbach equation (Thrailkill 1968a):

$$u^2 = (2dg/f) \cdot (dh/dl) = Q/a^2 \tag{2.9}$$

where f is a friction factor, from which

$$Q = \sqrt{(2dga^2/f)} \cdot \sqrt{(dh/dl)} \tag{2.10}$$

The relationship between f and Q has been studied by Lauritzen et al. (1985) for an active phreatic conduit in Norway. Apparent friction shows a rapid, exponential decrease with increasing discharge until a constant value is reached. Friction is affected by tube dimensions, wall roughness, and the complexity of conduit geometry, including the presence of breakdown.

Flow nets in karst drainage systems

The location of groundwater flow in a karst aquifer has been the subject of intense scientific debate since the time of Martel, and has been summarized by White (1988). Early workers considered that most flow occurred in the vadose zone rather than in the phreatic zone, but the latter view eventually predominated. Subsequently there was debate as to whether the locus of flow was close to the water table (the shallow phreatic theory of Swinnerton (1932), Rhoades and Sinacori (1941) and Thrailkill (1968a)) or some distance below it (the deep phreatic theory of Davis (1930) and Bretz (1942)). These various authors all tried to create a universal scheme applicable to all caves but based on the characteristics of the individual region that they studied. Subsequent approaches (Ford and Ewers 1978; Dreybrodt 1990) have used stochastic approaches to flow path location.

Most, if not all, of these approaches have been based on analysis of the morphology and context of drained conduits or caves. These are now realized to be a small fraction of the total conduit length, and the sample afforded by these has itself increased fourfold in the last twenty years (Courbon et al. 1989). Detailed work examining the relationships between passage morphology and geological structure (Ford 1965; Palmer 1987; Maire 1990) has added greatly to our understanding of why caves form where they do. The exploration and mapping of flooded caves by cave divers (see figure 2.4)

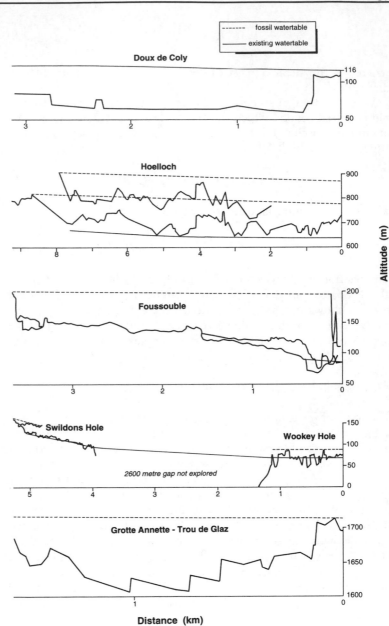

Figure 2.4
*Longitudinal sections of
flow paths in limestone
cave conduits, indicating
depth of circulation below
the water table. Conduits
in the Doux de Couly and
Swildons–Wookey systems
have been mapped by
divers; the remainder are
air filled and indicate fossil
water tables. From
Worthington (1991).*

has yielded even more useful information about active conduits up
to 6 km long (Grodzicki 1985; Palmer 1986).

The Role of Salinity

Where karst intersects the coast then the water table declines to
mean sea level. In many areas borehole data show that there is a

lens of freshwater overlying brackish to saline water. The thickness of the freshwater lens varies and may be significantly reduced by pumping, leading to saltwater intrusion and dominance. This phenomenon was first described by Ghyben and Herzberg, who found that the depth to the fresh/saltwater interface depends on the elevation of the water table above sea level and the densities of the waters.

$$Z_s = \frac{\rho f}{\rho_s - \rho_f} \cdot h_f \qquad (2.11)$$

where Z_s is the depth to the interface, ρ_f is the density of freshwater, ρ_s is the density of saltwater, and h_f is the elevation of the water table above sea level (see figure 2.5). If r_f is 1.0 and r_s is 1.025, then a 1 m change in the water table will cause a rise in the saline water of 40 m. This emphasizes the risk of unregulated pumping on saltwater intrusion. On the Pacific island of Nauru, a raised dolomitic limestone atoll (see figure 2.6), the dynamics of the interface have been investigated by Jankowski and Jacobson (1991). Nauru has been extensively mined for phosphatic fertilizer and its population is dependent on groundwater in this relatively dry part of the Pacific. Under Nauru large-scale drilling has revealed a thin freshwater lens overlying a thick mixing zone with a strong gradient of salinity to seawater at a depth of around 70 m (see figure 2.6). Groundwater flow is radially outwards to the sea and recharge is from vadose water percolating through the karst as well as conduit flow in caves. The freshwater layer is a chemically closed system whose chemistry is controlled by ingassing of carbon dioxide and by dissolution and precipitation reactions. In contrast the vadose waters, cave water and mixing zone are open systems where chemical evolution is controlled by mixing with seawater. Active karst dissolution is occurring at depth. Further extraction of freshwater may provoke a dramatic rise in the mixing zone yielding unpotable water. Drinking water is already being imported to Nauru from Australia at $10 per kilolitre.

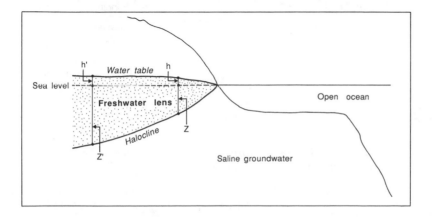

Figure 2.5
The depth Z to the fresh / saltwater interface (the Gyben–Herzberg interface) depends on the elevation h of the water table above sea level and the density of the waters.

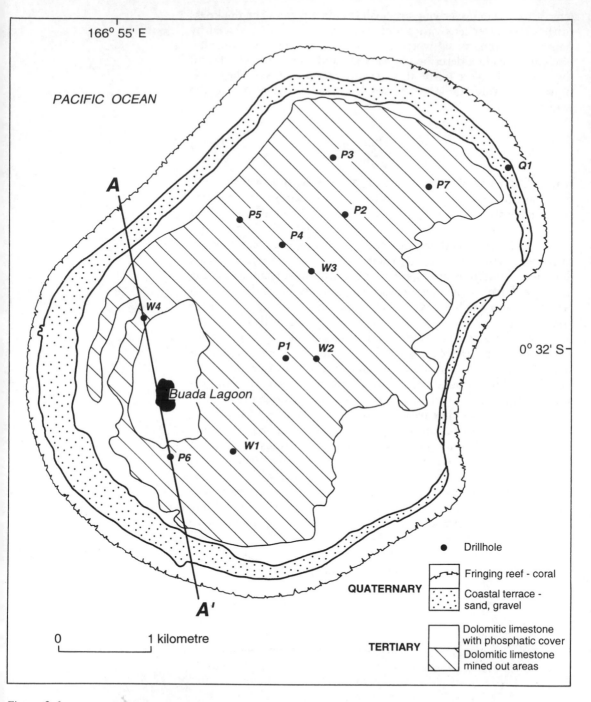

Figure 2.6
Plan and section of Nauru Island, Central Pacific, showing its surficial geology, the zonation of groundwater and its conductivity gradient. From Jankowski and Jacobson (1991).

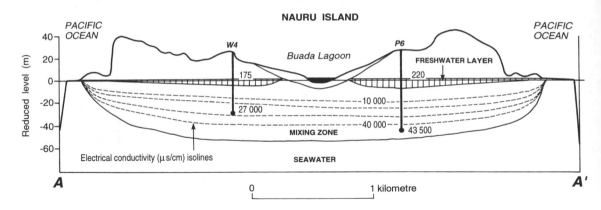

Evolution of the Karst Drainage System

Figure 2.6 *Continued*

From the previous section it is apparent that the location of the flow net below the karst water level is dependent on the rock structure and the length of the drainage system from inputs to outputs. Most phreatic conduits contain a single loop in the saturated zone, but multiple loops are known where there are local peculiarities of structure or in steeply bedded limestones with interbeds. Below the current active loop or tier is a complex of developing conduits which Worthington (1991) has argued are dominated by sulphate-rich waters. These conduits may also form along other chemoclines (fresh–salt) or along what Lowe (1989) has termed inception horizons, zones of mineralization or unconformities. These will be explored further in the next chapter.

There is some evidence to suggest that successive tiers of phreatic conduits are equally spaced in line with a Hagen–Poseuille flow net (Worthington 1991). An individual tier may be active for 0.3 to 10 million years (Worthington 1991, 130) based on radiometric and palaeomagnetic dating, but eventually it will be abandoned owing to intrinsic changes such as blockage by sediments or extrinsic changes such as base level lowering or spring aggradation. Thus any conduit may be envisaged as passing through a number of phases during its long life (White 1988; see figure 2.7). During the relatively short initiation phase, conduits are of small diameter and experience the transition from laminar to turbulent flow. Active enlargement to tubes of 3 to 5 m diameter occurs over a notional timespan of up to 100 000 years. After this flooded conduits may enlarge until they collapse and other tiers take over, or may be drained. If base level lowering is active then vadose canyons may result, or dry tubes may persist until the unsupported passages fragment or are removed by surface erosion. The likely timespan from initiation to decay is of the order of 1 to 10 million years. Small infilled fragments of these conduits may persist in the geological record as palaeokarst. In the next chapter the development of

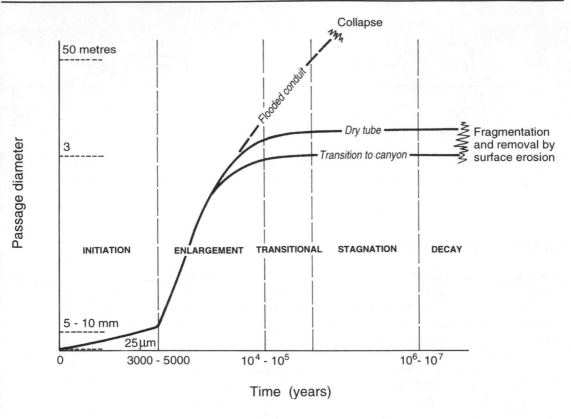

Figure 2.7
Schematic evolution of a
single karst conduit
through critical thresholds
separating flow regimes.
From White (1988).

individual conduits or caves in response to the solutional process
and geology is outlined.

Analysis of Karst Drainage Systems

Water tracing techniques

The technique of water tracing using artificial dyes or particles is
very widely employed in groundwater hydrology and is the subject
of several major works (Aley and Fletcher 1976; Milanovic 1981;
Quinlan 1987; Hotzl and Werner 1992; Fetter 1988). The most
common use has been the identification of flow routes from inputs
to output springs (see plate 2.4) or seepages. Repeated tracing of
flow routes has shown that there is a strong relationship between
discharge and tracer travel time (Smart 1981; Stanton and Smart
1981), and that flow paths may vary with stage as alternative routes
become activated. It is therefore necessary to carry out dye-tracing
experiments over the full range of discharges from low to high, and
to monitor all possible springs, however unlikely the connection
may seem to be.

Plate 2.4
Flood discharge from the
Allt nan Uamh spring,
Inchnadamph, Scotland.
Photo by Gareth Davies.

A number of substances have been used for tracing experiments. These include microorganisms, environmental isotopes, salts, dyes and spores, and they are fully discussed in Ford and Williams (1989, 219). Because of their low cost, low toxicity, easy detectability and high sensitivity, fluorescent dyes have been and will continue to be the most widely utilized in karst hydrology. The application of precision fluorometric techniques permit a definition and estimation of flow networks and hydraulics that are not readily

Table 2.2 Fluorescent dyes used in water tracing (modified from Ford and Williams 1989, 232–3)

Fluorescent dye	Colour index general name & number	Peak excitation nm	Peak emission nm	Minimum detectability μgL−1	pH effect	Adsorptive dye loss	Photochemical decay rate	Background fluorescence interference
Leukophor C	Fluorescent brightener 231	349	442	0.36	emission reduced below pH7	high on organics	very high	domestic detergents, organic leachates
Tinopal CBS-X	Fluorescent brightener	355	430	0.36	emission reduced below pH7	high on organics	very high	domestic detergents, organic leachates
Amino G acia (blue)		355	445	0.51	emission maximum at pH6–8.5	low	high	?
Fluorescein LT (green)	Acid yellow 73 (CI 45350)	490	520	0.29	emission reduced below pH7 to trough at pH3	moderate	high	organic leachates
Lissamine FF (green)	Acid yellow 7 (CI 56025)	420	515	0.29	emission maximum at pH4–10	low	low	organic leachates
Rhodamine WT (orange)	Acid red 388	555	580	0.10	emission reduced below pH5.5	moderate	low	none significant

available from other techniques. The principal dyes in use are outlined in table 2.2. Smart (1984) has provided a very thorough review of the toxicity of commonly used fluorescent dyes: only Tinopal CBS-X (an optical brightener), Fluorescein and Rhodamine WT have no demonstrated carconogenic or mutagenic hazard.

Two important conditions must be met for artificial dye tracers to yield reliable results. First, the tracer must be conservative, that is it must exhibit little or no loss within the karst drainage system. This is unfortunately not tenable due to the adsorption of fluorescent dyes onto clay-rich sediments or organic matter. Secondly, it is important that the tracer signature is unique or unequivocal. It must be clearly separable from other artificial dyes (see figure 2.8) and it must be distinct from naturally occurring compounds. Failure to meet these two criteria is often the cause of a false interpretation of a dye-tracing experiment.

There are three specific problems that need to be considered here: suppression, adsorption and natural fluorescence. The presence of certain ions may cause a suppression or quenching of fluorescence (Smart and Laidlaw 1977). Large losses of Rhodamine WT fluorescence were noted in the presence of sodium and potassium chloride, with loss increasing through time. However, the study of Bencala et al. (1983) showed little suppression in solutions containing lithium, sodium, potassium, strontium, chloride and nitrate. Dye adsorption is a more serious problem, although many studies assume that the tracer is conservative. Rhodamine WT has long been perceived as being conservative, though major losses can be expected in a range of hydrological environments. Trudgill (1987) investigated adsorption in soils, while Bencala et al. (1983) recovered only 45 per cent of the Rhodamine after a flow length of 300 m over sediment. Fluorescein is readily adsorbed by clays and also probably by colloidal organic matter.

Natural fluorescence in soil and runoff water has been well documented (Ghassemi and Christian 1968; Smart and Laidlaw 1977; Trudgill 1987). It varies with time, flow conditions and source areas. Increased fluorescence is often noted under high flow conditions owing to flushing of organic compounds from the soil. Fluorescence spectra for natural waters have broad band width in the blue-green wavelengths, and this is largely owing to dissolved organic matter. The natural orange-red band (due to phytoplankton) is many times smaller than the blue-green band, and this in part explains the preference for Rhodamine over Fluorescein. Although natural fluorescence can be separated from that of dyes in the laboratory using a spectrofluorometer, this is not yet possible in the field.

The form of the time-concentration or tracer breakthrough curve depends on the structure of the conduit network, the prevailing flow conditions and the nature of the tracer used (Smart 1988b). Consideration of these factors in combination with the regional

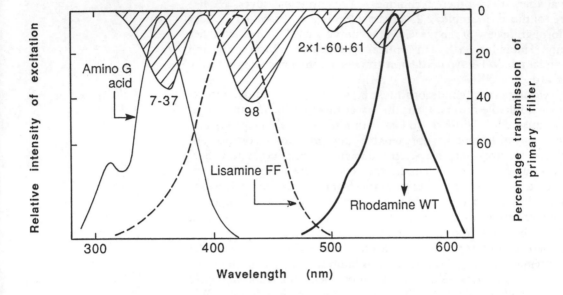

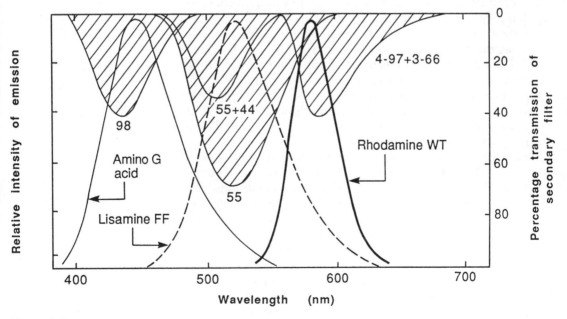

Figure 2.8
Absorption and emission spectra of blue (Amino G acid), green (Lissamine FF) and orange (Rhodamine WT) fluorescent dyes. From Smart and Laidlaw (1977).

hydrogeology can yield a great deal of information about the structure and dynamics of the karst drainage system (e.g., Smart 1988a; Crawford 1994). The form of the breakthrough curve has been used to indicate off-line storage in the conduit network (Atkinson et al. 1973; Smart 1983).

Spores of the club moss *Lycopodium clavatum* were once widely employed for dye tracing, but their use has declined since the 1970s. The best overall reviews of their use are Drew and Smith (1969) and Smart and Smith (1976). The spores have a diameter of 30–35µm and are slightly buoyant in freshwater. They can drift though karst conduits with little filtration and be recovered using plankton nets with a 25µm mesh. The principal advantage of spores for tracing is that they can be dyed in up to six colours and thus employed for tracing multiple inputs to a karst drainage system, where use of about three dyes is the maximum practicable. The main disadvantages of spores are that operator error in colour identification may be considerable, that the aquifer is contaminated with spores for a long time, and that the technique is non-quantitative because the trap efficiency of the nets and laboratory recovery rates are unknown (Smart et al. 1986).

Environmental isotopic techniques are based on measurements of variation on the isotopic composition of natural waters, and are most useful in problems related to the dynamics of drainage systems (using the radioactive isotopes tritium and carbon-14) and the origin of waters (using the stable isotopes deuterium and oxygen-18) (Bradley et al. 1972; IAEA 1983). In Crete, Leontiadis et al. (1988) determined the origin of recharge to coastal springs from karst plateaux, and estimated the age of groundwater to range from 45 to >100 years using tritium input functions. The high variability in percolation input to a shallow cave in Israel was noted by Even et al. (1986). Using tritium they determined the age of seepage water to be several decades. Enrichment of deuterium and oxygen-18 in the seepage water suggested that the water was derived from a relatively dry winter in 1962–3 when westerly air masses dominated the region. In Australia Allison et al. (1985) have investigated the rate of recharge of groundwater in calcretes using stable isotopes. Chlorine-37 concentrations in soil can increase dramatically through evaporation, and this saline water moves slowly to the groundwater table and hence to rivers. It is clear that, where the mallee vegetation has been cleared, recharge rates are increased. Recharge rates vary from 60 mm yr^{-1} for secondary dolines to 0.06 mm yr^{-1} for calcrete flats. Recharge rates for vegetated dunes are 0.06 mm yr^{-1}; for cleared dunes they are 14 mm yr^{-1}. Thus the highest rates of saline recharge are associated with secondary collapse dolines, where hydraulic conductivity is enhanced by fissures. There is considerable scope for further investigation of the authigenic percolation system in karst using these techniques.

Figure 2.9
*Streamsink and spring
flood hydrographs from
Takaka Hill, NewZealand,
showing lag effects and
peak flattening. From
Williams (1984).*

Spring hydrograph analysis

Karst drainage basins respond to rapid changes in recharge in a
similar manner to surface basins. Analysis of the flood hydrographs
from a spring can reveal much about the structure and function of
the network feeding it. The most useful data come from monitoring

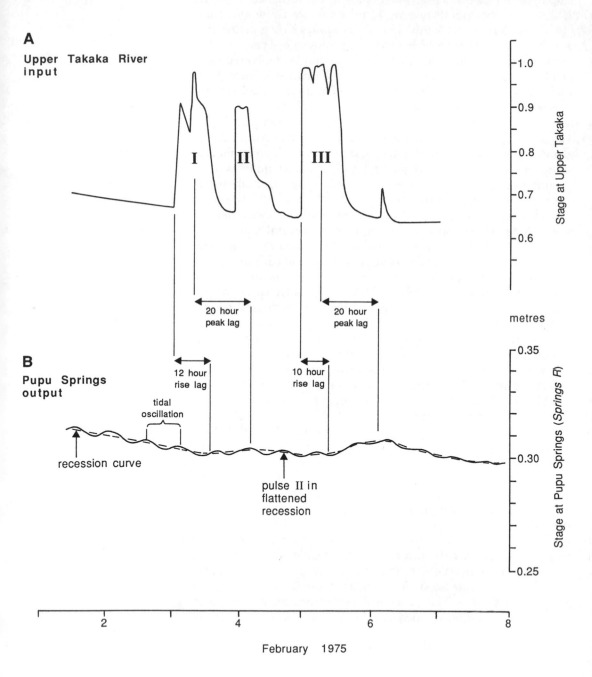

A

**Upper Takaka River
input**

B

**Pupu Springs
output**

February 1975

of intense storms that inject pulses of water into the karst. Streamsink hydrographs have a steep rising limb as overland flow feeds the input point, a peak flow (Q_{max}) and a long declining limb or recession to the baseflow, unless another storm occurs (see figure 2.9). There is usually a lag time before the pulse is detected as a rise in discharge at the spring. The length of this lag will depend on the structure of the karst drainage system and the amount of stored water in the system. Thus for a fairly direct flow path, with little stored water, most of the inflowing water will be stored temporarily in fissures and conduits, and the rise at the spring will be slow and of modest size. There is usually considerable damping of the spring hydrograph, with a broad peak in discharge. In contrast, if antecedent rainfall has filled the available conduit and fissure storages, then the pulse will be rapidly transmitted with a small lag, and the spring hydrograph response may be dramatic and the peak may resemble the inflow hydrograph with little damping. Often there will be flushing of older, stored water from the conduits as the inflowing water acts like a piston.

If the spring is monitored for water chemistry and sediment load as well as discharge, then this flushing effect may be detected and the size of the storage estimated (Ashton 1966). An idealized spring hydrograph is shown in figure 2.10. The hydrograph can be sepa-

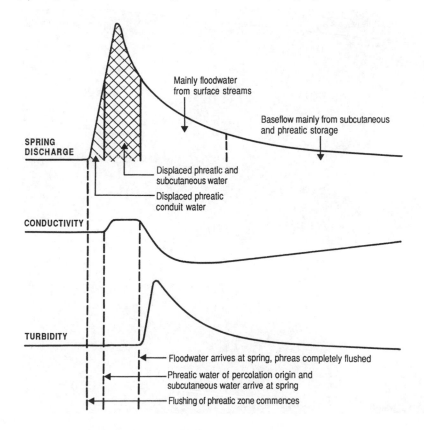

Figure 2.10
Interpretation of an idealized spring hydrograph and chemograph. From Williams (1983).

rated into zones where a particular water source is dominant. On the initial rising limb, phreatic conduit water of long residence time may be displaced by incoming water. This is replaced by a mix of phreatic and subcutaneous water as the main flood pulse comes through, characterized by higher conductivity and increasing turbidity. Finally surface floodwater with a great deal of suspended sediment (see plate 2.5) raises turbidity but has lower solute concentration (lower conductivity). This water dominates the early part of the recession but is supplanted by baseflow from subcutaneous and phreatic storage. The magnitude of each water type can be roughly estimated by integrating under the separated hydrograph. This approach also permits some clarification of conduit network geometry by interpretation in terms of combinations of spatially distributed inputs to produce complex output pulses (see figure 2.11). Since Ashton's theoretical work, several authors have considered pulse wave analysis purely in terms of discharge (Brown 1972, 1973; Williams 1977), while others have combined discharge with water chemistry in pulse analysis (Christopher 1980; White 1988). An elaboration of the technique is the labelling of natural or artificial pulse waves with fluorescent dyes (Christopher et al. 1981; Smart and Hodge 1980).

Plate 2.5
Active vadose incision is proceeding apace in the entrance chamber of Atea Kananda Cave, Papua New Guinea. Below the figure the river flow of $4\,m^3\,s^{-1}$ enters a slot $3\,m$ wide.

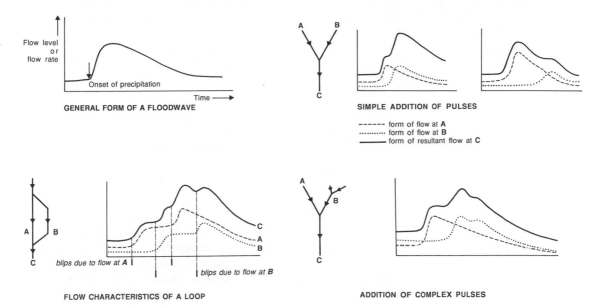

GENERAL FORM OF A FLOODWAVE

SIMPLE ADDITION OF PULSES

----- form of flow at **A**
.... form of flow at **B**
—— form of resultant flow at **C**

FLOW CHARACTERISTICS OF A LOOP

ADDITION OF COMPLEX PULSES

Springs with low variance in flow and chemistry are often fed solely by autogenic recharge, that is water derived wholly from the epikarst. In contrast, springs with high variance in flow and chemistry are often associated with allogenic recharge (Jakucs 1959), where water is derived partially or wholly from non-karstic rocks. Rarely is a spring fed by an entire karst drainage system and can be termed a full-flow spring (Worthington 1991). More commonly there is a hierarchy of springs draining a karst with underground distributaries feeding springs at differing levels in a valley. The lowest elements preferentially fed by baseflow are termed underflow springs, while upper springs preferentially fed by flood flow are called overflow springs. There are intermediate members in the hierarchy called overflow-underflow springs. Smart (1988d) studied about eighty springs over a 150m vertical range near Castleguard Cave, Alberta, Canada. Steady and sustained flow came from the lowest underflow members, while short-lived variable flow came from the upper members, with the uppermost fossil springs showing no flow. Underflow springs may be hard to detect as they may be aggraded by alluvium or sub-lacustrine or be seepages in a stream bed.

Where detailed discharge records are available, underflow and overflow springs may be distinguished by the ratio of maximum annual discharge to minimum annual discharge and the proportion of time that flow is present (see table 2.3).

The shape of the flood hydrograph recession is a good diagnostic tool (Atkinson 1977; Mangin 1975). Discharge usually decreases exponentially during the recession to baseflow such that

Figure 2.11
Flood pulse generation and hydrograph forms resulting from flow addition and complex flow paths. From Ashton (1966).

Table 2.3 Discharge characteristics for distinguishing spring flow types (from Worthington 1991, 51)

Spring type	Qx/Qn	Days with Q > 0
Full flow	high	all
Underflow	low	all
Overflow	∞	few–all
Underflow-overflow	∞	few–all

Q_x is maximum annual discharge while Q_n is minimum annual discharge.

$$Q_t = Q_0 \, e^{-\alpha t} \qquad (2.12)$$

where Q_0 is the baseflow at the start of the recession (t = 0) and α is the recession exponent. Q is usually expressed in $m^3 s^{-1}$ while t is expressed in days. There may be seasonal variation in the recession exponent depending on recharge and storage.

The shape of the recession can be used to diagnose the spring type (Worthington 1991, 52). For full flow springs, α is constant (see figure 2.12). This is relatively rare, though many karst studies have assumed that it obtains: in the Mammoth Cave area, Quinlan and Ewers (1989) found that only one of the 21 larger karst drainage systems was drained by a full flow spring.

Overflow springs are characterized by convex log-normal baseflow recession (α increasing with time), with strongly seasonal or intermittent springs having a minimum discharge of zero and α tending to ∞. Underflow springs may be of two types. Losing or high-stage underflow springs are controlled by constrictions or aggradation close to the output, so at high stage the water backs up in the conduit until an overflow outlet is activated. This provides a constant head and a slightly decreasing or constant α depending on the magnitude of the recharge from the catchment relative to the output. Gaining or low-stage underflow springs have a concave recession, with α decreasing as a full-flow recession is supplemented by underflow from surface stream beds or other cave streams. Finally, many karst springs fall into the last category of underflow-overflow, where the spring is intermediate in a hierarchy and functions as an underflow spring for part of the time and then reverts to the intermittent overflow type.

In most karst drainage systems each overflow spring has a complementary underflow spring. There is also some loss or gain from surface streams to groundwater in any karst. Finally, some groundwater flow will always emerge at the lowest point in the karst drainage system owing to the hydraulic head, but may not do so as a discrete spring. Expect the unexpected!

Spring chemograph analysis

Where a spring has been monitored for both discharge and conductivity, it is possible to examine the seasonal and annual variation in

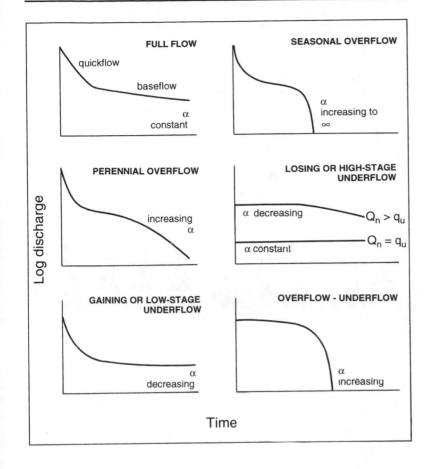

Figure 2.12
*Karst spring types in
relation to the exponent α
of the recession curve.
From Worthington (1991).*

solute load from the karst drainage system. There is usually a very
good statistical relationship between conductivity and total hardness, and this can be used to calculate the solute load. Figure 2.13
is the ten-year record from the Argens spring in Provence, southern
France (Julian and Nicod 1989). This large underflow spring drains
the karst plateau to the north of Ste Victoire and shows strongly
seasonal flow, with a maximum in winter. This is typical of karst
springs whose recharge is due to snow melt. There is a strong
relationship between mean monthly discharge and solute load, with
baseflow making a significant contribution in all months. Low
recharge during the 1967–8 drought was compensated by inter-
annual carry over of karst water.

The frequency distribution of conductivity from such springs can
be used to characterize their aquifers (Bakalowicz and Mangin
1980):

> Diffuse or porous aquifers – unimodal, high conductivity
> Fissured aquifers – unimodal, low conductivity
> Conduit or karstified aquifers – polymodal, wide range of
>   conductivity.

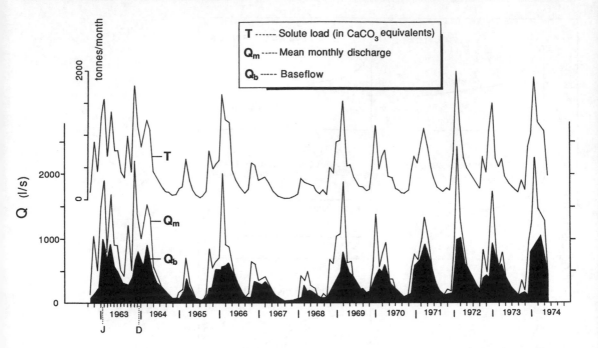

Figure 2.13
*Ten-year spring
hydrograph and
chemograph from the
Argens spring, Provence,
France. Note the
seasonality of flow, the
large proportion of
baseflow discharge, and
the dependence of solute
load on flood discharges.
From Julian and Nicod
(1989).*

Thus in figure 2.14 the Evian mineral water source is of the diffuse type, while the Source de Surgeint aquifer is probably more fissured. Most of the other examples are of the mixed conduit type, with a number of populations of unique geochemical evolution contributing to the final histogram.

Structure and Function of Karst Drainage Systems

Storage and transfers in the karst system

One consequence of the model of Smart and Hobbs (1986) is that spring hydrograph response can be affected by combinations of recharge, storage and flow processes. In figure 2.15, the effects of changing these factors is explored. Concentrated recharge (for example, snowmelt) into a karst drainage system with low storage will produce a peaked spring hydrograph, while if storage is high the hydrograph will be flattened as water is gradually released. If diffuse flow is present as well, then the hydrograph is attenuated twice to produce a broad spring hydrograph more suggestive of baseflow dominance (see plate 2.6). With a dispersed recharge due to autogenic recharge in the epikarst, the nature of storage and the presence or absence of diffuse flow will radically alter the spring hydrograph to produce a wide range of responses. This complexity indicates that analysis of spring hydrographs alone is unlikely to

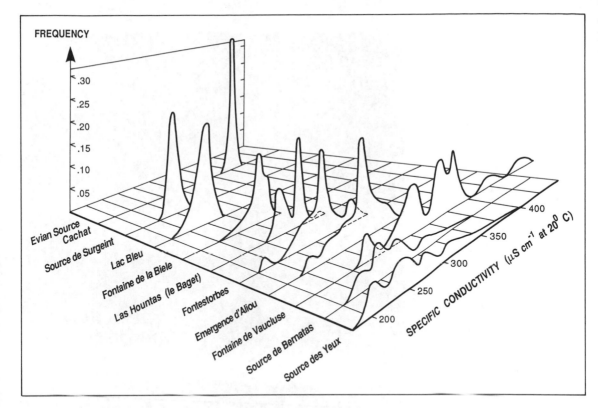

Figure 2.14
Frequency distributions of conductivity of karst spring waters in southern France. From Bakalowicz and Mangin (1980).

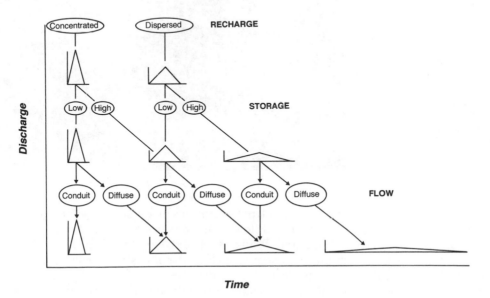

Figure 2.15
Effects of variation in recharge, storage and flow types on the form of flood hydrographs in karst drainage systems. From Smart and Hobbs (1986).

Plate 2.6
The resurgence of Longgong Dong (Dragon Palace Cave), Guizhou, China, is harnessed for a hydroelectric plant, as well as being used for tourism. Inside the cave the water from a 600m long lake falls 36m to the entrance.

produce sound interpretations unless some knowledge of the nature of storage and flow characteristics can be obtained.

Ford and Williams (1989) have developed a model of karst drainage systems which incorporates these factors and serves as a basis for enhanced understanding of karst hydrology (see figure 2.16). If the major storages and fluxes in this scheme can be quantified, then the response of karst drainage systems to recharge events and to human-induced changes can be better estimated. Key

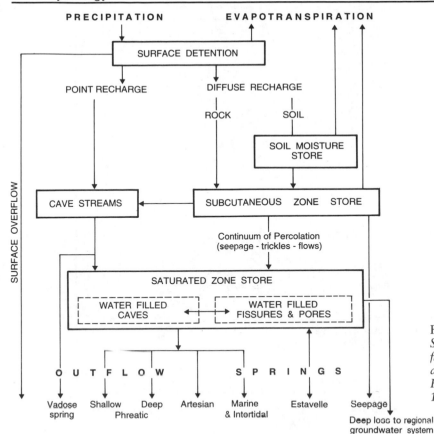

Figure 2.16
*Scheme of storages and
flow linkages in a karst
drainage system. From
Ford and Williams (1989,
169).*

elements of this scheme which require better understanding are the
subcutaneous zone store with its included percolation system
(Friederich and Smart 1982), and the saturated zone store with the
interchange between conduit (cave) storage and the fissure storage.
The latter will require considerable input from cave diving scientists, but significant progress in this area has been made (Lauritzen
et al. 1985; Palmer 1986; Worthington 1991, 1994).

The role of extreme events

The structure of a karst drainage system is dynamic in that certain
linkages may be activated only once a limit is passed on a rising
stage. A good example is provided by the Parker Cave system of
Kentucky (Quinlan et al. 1991), which has five entry points to a
9 km long cave conduit (see figure 2.17). At baseflow conditions
each of the five conduits carries a small flow. If stream 2 floods,
then underground flow occurs in two directions, while if stream 1
floods the flow is unidirectional because of relative passage levels.
However, a lagged stage rise in stream 1 will also produce bi-
directional flow, and a series of rain storms will cause complete
flooding of all five conduits and multi-directional flow. This com-

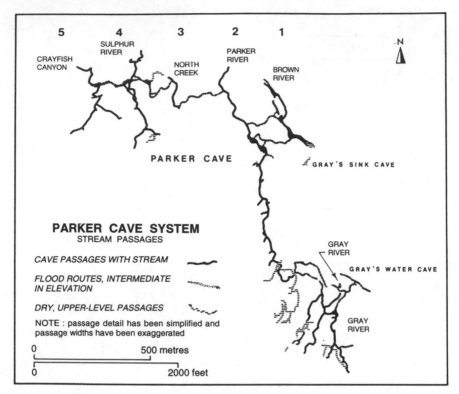

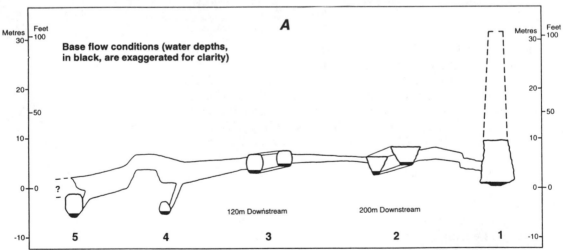

Figure 2.17
Flow paths in Parker Cave, Kentucky, following rainfall events of different magnitudes. Numbered passages in the plan refer to passage cross-sections in A (baseflow) and B–D (varying flood flows). From Quinlan et al. (1991).

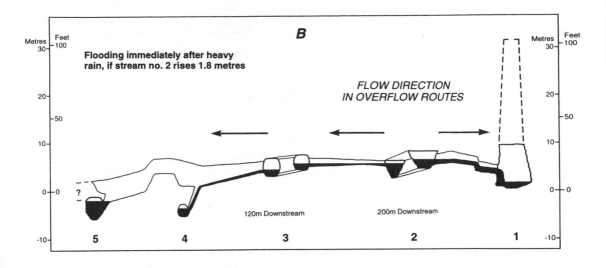

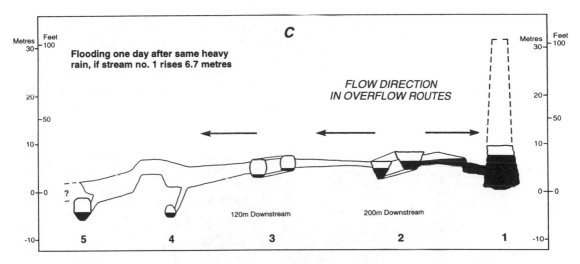

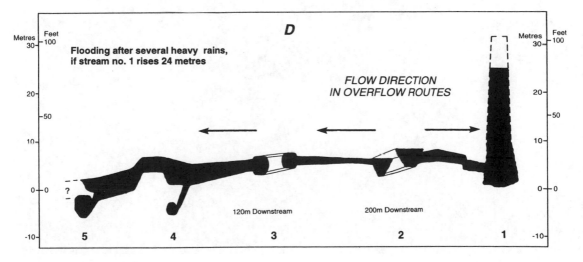

Figure 2.17 *Continued*

plexity of flow, and dependence on antecedent rainfall, is typical of karst and creates unpredictability when pollutants become mobilized.

The nature of underflow-overflow springs in a hierarchy has already been described, and both Smart (1988c) and Brown (1972) have provided examples of the dynamism of the conduit networks from the Canadian Rockies. The extent to which fissure storage is filled by a single recharge pulse will determine the nature of the spring hydrograph, and there may be variable lags between the input pulse and the response at the spring. When we consider the response of the karst drainage system to extreme events, such as the 1 in 100-year flood event, the prediction of behaviour is far less certain.

Each distributary link in a network will have its own unique characteristics of hydraulics and storage, and their combination may be variable in time and space. The friction within each link will probably decrease with increasing discharge. In this context the role of hydraulic semiconductors, described previously, becomes crucial to our understanding. Intrinsic changes in the nature of the conduit, due to the mobilization of sediment banks and destabilization of breakdown, may occur at long intervals during exceptional floods. This may explain the sudden backing up of water in caves (see plate 2.7) so that vadose conduits become temporary phreatic for periods

Plate 2.7
The streamsink of the Tekin River cave, Oksapmin, Papua New Guinea, shows large quantities of sediment and woody debris transported to the insurgence. When blocked this sink has a lake 10 m deep at the cave entrance.

of days to weeks, and active enlargement by solution may occur. This has been observed in the large caves of New Guinea by Checkley (1993) for the large sink of the Baliem River in Irian Jaya, and by Francis et al. (1980) for the passages of Atea Kananda in the Muller Range. Palaeokarstic passages may become active and change the boundary of the karst drainage system (Kiernan 1993).

The radical changes in water level in the poljes of the Nahanni karst, northern Canada, have been documented by Brook and Ford (1980). Although the spring thaw may be the annual hydrologic event of greatest magnitude in most of the arctic, in the Nahanni intense summer storms have profound effects. After 203 mm of rain over a week, the poljes filled as marginal sinks were incapable of coping with the inflow. At Raven Lake alone the water level rose by 49 m. Alluviation of the poljes is held to account for the periodic flooding, and although the karst system can cope with annual snowmelt runoff, low frequency spring and summer storms may have a greater role in karst development in the region than was hitherto realized.

Karst Hydrology of the Mammoth Cave Plateau, Kentucky

Underlying the gently rolling farmland and forests of central Kentucky are some of the longest and most interesting caves in the world. Over 800 km of cave passages form the Mammoth Cave system, part of which lies in a national park of the same name. Most of the caves are formed in the Girkin, Ste Genevieve and St Louis limestone formations, which dip gently to the northwest and are wrinkled by a number of small anticlines. The limestone is in places overlain by insoluble sandstones and shales. To the south of Mammoth Cave National Park these rocks have been largely eroded away to form the Pennyroyal Plateau, and limestone occurs just under the soil over a very large area punctuated by many sinkholes. To the north these resistant rocks form the dissected ridges and small cuestas of the Chester Upland, with limestone in the deep valleys. Groundwater in the limestone is derived in part from drainage from these high areas, and there are numerous sinkholes along the edge of the sandstone uplands. The Green River lies in a deep trench and is fed by many springs from the limestone as it flows west to join the Ohio River.

The area is thus underlain by cavernous limestone and contains thousands of dolines, about a hundred streamsinks and about two hundred springs. Generally, all surface and subterranean water within a groundwater basin flows to the same spring or set of springs. Water or pollutants from a point source can be dispersed through distributaries within the karst drainage system to as many as 53 springs along the Green River in the Bear Wallow groundwater basin. This is an extreme example, most being fewer

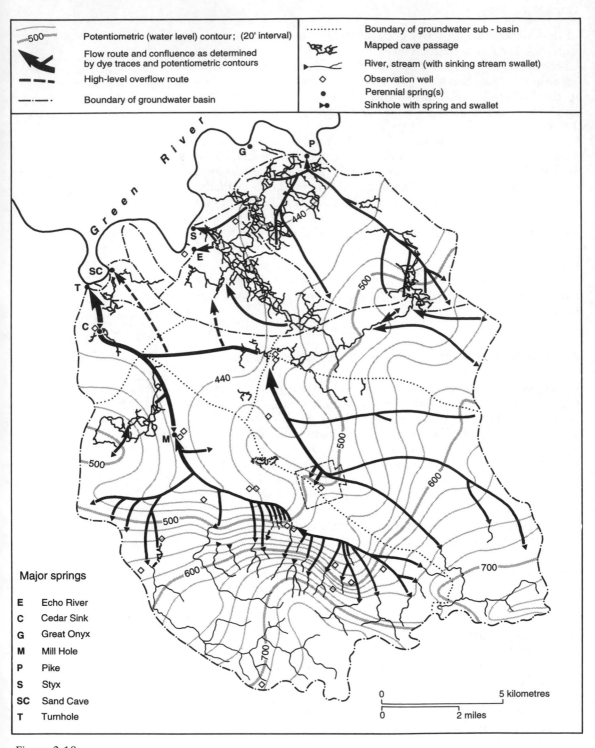

Legend:

- ~~~500~~~ Potentiometric (water level) contour; (20' interval)
- Flow route and confluence as determined by dye traces and potentiometric contours
- High-level overflow route
- Boundary of groundwater basin
- Boundary of groundwater sub - basin
- Mapped cave passage
- River, stream (with sinking stream swallet)
- ◇ Observation well
- ● Perennial spring(s)
- ►●► Sinkhole with spring and swallet

Major springs

E	Echo River
C	Cedar Sink
G	Great Onyx
M	Mill Hole
P	Pike
S	Styx
SC	Sand Cave
T	Turnhole

0 ___ 5 kilometres
0 ___ 2 miles

Figure 2.18
Hydrology, potentiometric surface and underground flow routes defined by dye tracing in the Turnhole Spring karst basin of Mammoth Cave, Kentucky. From Quinlan et al. (1991).

in number, but the Mammoth Cave area is typical of the highly
interconnected nature of karst drainage systems. The area is heavily
used for agriculture, for grazing and for tourism, and there is thus
a wide range of potential pollutants for karst groundwater. Fortunately this karst is quite well known and has been the subject of
many scientific studies.

Our detailed knowledge of the hydrology of the Mammoth Cave
region is the result of intensive study over the last two decades by
Jim Quinlan, Ralph Ewers, Joe Ray and other workers. This work
has entailed more than five hundred dye-tracing experiments, 2700
water level measurements, and mapping of more than 800 km of

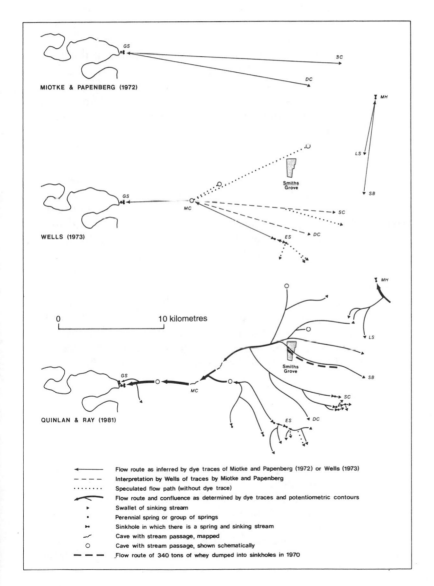

Figure 2.19
*The evolution of
knowledge about the
Graham Springs karst
basin, Kentucky, based on
dye-tracing experiments:
DC, Doty Creek; ES, Elk
Spring; GS, Graham
Springs; LS, Little Sinking
Creek; MC, Mill Cave;
MH, Mill Hole; SB,
Sinking Branch; SC,
Sinking Creek. From
Quinlan et al. (1991).*

cave passages by the Cave Research Foundation, National Parks Service and other speleological organizations. It is probably the most intensively investigated karst in the world. A detailed map of flow routes and the potentiometric surface (Quinlan and Ray 1981) is but one of the many products of this research. Any serious student of karst hydrology should study this map as an example of how to conduct a detailed investigation.

In figure 2.18 the detailed hydrology of part of this map is shown for the Turnhole Springs groundwater basin. This includes Mammoth Cave and Roppel Cave in the northeast of the map, and Park City urban area in the centre. The flow of water is generally orthogonal to the water-level contours, and there is concentration of flow along troughs in the water-level contours. These troughs coincide with major conduits or caves, such as the Mill Hole or Roppel Cave conduit networks.

The evolution of understanding about flow paths in this basin is illustrated in figure 2.19. Early dye-tracing work by Miotke and Papenberg indicated that water flowed directly from Sinking Creek (SC) and Doty Creek (DC) to Graham Spring (GS). Later work by Wells indicated that these flow paths actually went through Mill Cave (MC) and were joined by other caves to the north and flow from the Elk Spring (ES) complex. In addition, surface water from close to Sinking Creek entered two streamsinks (Sinking Branch, SB, and Little Sinking Creek, LS) and emerged at the Mill Hole (MH) spring. The lowest diagram shows the complex pattern of groundwater flow mapped by Quinlan and Ray. Many more tributaries to the Graham Spring karst drainage system are evident, and there is a dense flow network which defines a groundwater basin boundary to the east of Smith's Grove. This divide is narrow, and under storm flow conditions water could be diverted to either Graham Springs or Mill Hole.

Under flood conditions many of these groundwater basin boundaries may be breached and flow may occur to other springs. This has important ramifications for the flow of pollutants which inadvertently enter the karst drainage system of the Mammoth Cave area. This topic is further explored in Chapter 9.

References

Aley, T. and Fletcher, M. W. 1976: Water tracers' cookbook. *Missouri Speleology* 16 (3), 1–32.

Allison, G. B., Stone, W. J. and Hughes, M. W. 1985: Recharge in karst and dune elements of a semi-arid landscape as indicated by natural isotopes and chloride. *J. Hydrol.* 76, 1–25.

Ashton, K. 1966: The analysis of flow data from karst drainage systems. *Trans. Cave Research Group G.B.* 7 (2), 161–203.

Atkinson, T. C. 1977: Diffuse flow and conduit flow in limestone terrain in the Mendip Hills, Somerset (Great Britain). *J., Hydrol.* 35, 93–110.

Atkinson, T. C. 1985: Present and future directions in karst hydrogeology. *Ann. Soc. Géol. Belgique* 108, 293–6.

Atkinson, T. C. and Smart, P. L. 1979: Traceurs artificiels en hydrogéologie. *Bulletin BGRM* (2e ser.) III (3), 365–380.

Atkinson, T. C., Smart, D. I., Lavis, J. J. and Whitaker, R.J. 1973: Experiments in tracing underground waters in limestones. *J. Hydrol.* 19, 323–49.

Bakalowicz, M. and Mangin, A. 1980: L'aquifère karstique: sa définition, ses characteristiques et son identification. *Mm. L. sr. Soc. Géol. France* 11, 71–9.

Bencala, K. E., Rathbun, R. E., Jackman, A. P., Kennedy, V. C., Zellweger, G. W. and Avanzino, R. J. 1983: Rhodamine WT losses in a mountain stream environment. *Water Resources Bulletin* 19, 943–50.

Bradley, E., Brown, R. M., Gonfiantini, R., Payne, B. R., Przewlocki, K., Sauzay, G., Yen, C. K. and Yurtsever, Y. 1972: Nuclear techniques in groundwater hydrology. In *Groundwater Studies: An International Guide for Research & Practice*. Paris: UNESCO, ch. 10.

Bretz, J. H. 1942: Vadose and phreatic features of limestone caves. *J. Geol.* 50 (6), 675–811.

Brook, G. A. and Ford, D. C. 1980: Hydrology of the Nahanni karst, northern Canada, and the importance of extreme summer storms. *J. Hydrol.* 46, 103–21.

Brown, M. C. 1972: *Karst Hydrology of the Lower Maligne Basin, Jasper, Alberta*. Cave Studies 13, Cave Res. Ass., CA, 178 pp.

Brown, M. C. 1973: Mass balance and spectral analysis applied to karst hydrologic networks. *Water Resources Research* 9, 749–52.

Checkley, D. 1993: Cave of Thunder: the exploration of the Baliem River cave, Irian Jaya, Indonesia. *International Caver* 6, 11–17.

Christopher, N. S. J. 1980: A preliminary flood pulse study of Russett Well, Derbyshire. *Trans. Brit. Cave Res. Assoc.* 7 (1), 1–12.

Christopher, N. S. J., Trudgill, S. T., Crabtree, R. W., Pickles, A. M. and Culshaw, S. M. 1981: A hydrological study of the Castleton area, Derbyshire. *Trans. Brit. Cave Res. Assoc.* 8 (4), 189–206.

Courbon, P., Chabert, C., Bosted, P. and Lindsley, K. 1989: *Atlas of the Great Caves of the World*. St Louis: Cave Books, 368 pp.

Crawford, S. 1994: Hydrology and Geomorphology of the Paparoa Karst, North Westland, New Zealand. Unpublished PhD thesis, University of Auckland, 240 pp.

Davis, W. M. 1930: Origin of limestone caverns. *Bull. Geol. Soc. Am.* 41, 475–628.

Drew, D. P. and Smith, D. I. 1969: Techniques for the tracing of subterranean water. *Brit. Geomorph. Res. Group, Tech. Bull.* 2, 36.

Dreybrodt, W. 1990: The role of dissolution kinetics in the development of karst aquifers in limestone: a model simulation of karst evolution. *J. Geol.* 98, 639–65.

Even, H., Carmi, I., Magaritz, M. and Gerson, R. 1986: Timing the transport of water through the upper vadose zone in a karstic system above a cave in Israel. *Earth Surface Processes and Landforms* 11, 181–91.

Fetter, C. W. 1988: *Applied Hydrogeology*, 2nd edn. New York: Merrill, 592 pp.

Ford, D. C. 1965: The origin of limestone caverns: a model from the central Mendip Hills, England. *NSS Bull.* 27, 109–32.

Ford, D. C. and Ewers, R. O. 1978: The development of limestone cave systems in the dimensions of length and depth. *Can. J. Earth Sci.* 15, 1783–98.

Ford, D. C. and Williams, P. W. 1989: *Karst Geomorphology and Hydrology*. London: Unwin Hyman, 601 pp.

Francis, G., Gillieson, D. S., James, J. M. and Montgomery, N. R. 1980: Underground geomorphology of the Muller Plateau. In J. M. James and H. J. Dyson (ed.) *Caves and Karst of the Muller Range*. Sydney: Speleological Research Council, 110–17.

Freeze, R. A. and Cherry, J. A. 1979: *Groundwater*. Englewood Cliffs, NJ: Prentice-Hall.

Friederich, H. and Smart, P. L. 1982: The classification of autogenic percolation waters in karst aquifers: a study in G. B. Cave, Mendip Hills, England. *Proc. Univ. Bristol Speleo. Soc.* 16 (2), 143–59.

Ghassemi, M. and Christian, R. F. 1968: Properties of the yellow organic acids of natural waters. *Limnology and Oceanography* 13, 583–97.

Grodzicki, J. 1985: Genesis of the Nullarbor Plains caves in Southern Australia. *Zeitschrift für Geomorphologie* 29 (1), 37–49.

Hagen, G. 1839: Über die Bewegung des Wassers in engen cylindrischen Rohren. *Poggendorff Annalen* 46, 423–42.

Hotzl, H. and Werner, A. 1992: *Tracer Hydrology*. Rotterdam: A. A. Balkema.

International Atomic Energy Agency 1983: *Guidebook on Nuclear Techniques in Hydrology*, Technical Reports Series No. 91. Vienna: IAEA.

Jakucs, L. 1959: Neue Methoden der Höhlenforschung in Ungarn und ihre Ergebnisse. *Die Höhle* 10, 88–98.

Jankowski, J. and Jacobson, G. 1991: Hydrochemistry of a groundwater–seawater mixing zone, Nauru Island, Central Pacific Ocean. *BMR J. Aust. Geol. Geophys.* 12, 51–64.

Julian, M. and Nicod, J. 1989: Les karsts des Alpes du Sud et de Provence. *Z. Geomorph. N.F. Suppl. Bd.* 75, 1–48.

Kiernan, K. 1993: The Exit Cave quarry: tracing waterflows and resource policy evolution. *Helictite* 31, 27–42.

Lauritzen, S. E., Abbott, J., Arnesen, R., Crossley, G., Grepperud, D., Ive, A. and Johnson, S. 1985: Morphology and hydraulics of an active phreatic conduit. *Cave Science* 12 (3), 139–46.

Leontiadis, I. L., Paynes, B. R. and Christodoulou, T. 1988: Isotope hydrology of the Aghios Nikolaos area of Crete, Greece. *J. Hydrol.* 98, 121–32.

Lowe, D. J. 1989: Limestones and caves of the Forest of Dean. In T. D. Ford (ed.) *Limestones and Caves of Wales*. Cambridge: Cambridge University Press, 106–16.

Maire, R. 1990: La haute montaigne calcaire. *Karstologia – Mémoires* 3, 731 pp.

Mangin, A. 1975: Contribution à l'étude hydrodynamique des aquifères karstiques. DES thesis, University of Dijon (*Ann. Spéléo.* 29 (3), 1974, 283–332; 29 (4), 1974, 495–601; 30 (1), 1975, 21–124).

Milanovic, P. T. 1981: *Karst Hydrogeology*. Littleton, CO: Water Resources Pubs.

Palmer, A. N. 1987: Cave levels and their interpretation. *NSS Bull.* 49, 50–66.

Palmer, R. J. 1986: Hydrology and speleogenesis beneath Andros Island. *Cave Science* 13 (1), 7–12.

Poseuille, J. M. L. 1846: Recherches expérimentales sur le mouvement des liquides dans les tubes de très petits diamètres. *Acad. Sci. Paris Mem. sav. étrang.* 9, 433–545.

Quinlan, J. F. (ed.) 1987: *Practical Karst Hydrogeology with Emphasis on Groundwater Monitoring.* Dublin, OH: National Well Water Association 6, E1–E26.

Quinlan, J. F. and Ewers, R. O. 1989: Subsurface drainage in the Mammoth Cave area. In W. B. White and E. L. White (eds) *Karst Hydrology: Concepts from the Mammoth Cave Area.* New York: Van Nostrand Reinhold, 65–103.

Quinlan, J. F. and Ray, J. A. 1981: *Groundwater Basins in the Mammoth Cave Region, Kentucky.* Friends of the Karst, Occ. Pub. 1.

Quinlan, J. F., Ewers, R. O. and Aley, T. 1991: *Practical Karst Hydrology, with Emphasis on Ground-Water Monitoring. Field Excursion: Hydrogeology of the Mammoth Cave Region, with Emphasis on Problems of Ground-Water Contamination.* Dublin, OH: National Ground Water Association, 93 pp.

Rhoades, R. and Sinacori, N. M. 1941: Patterns of groundwater flow and solution. *J. Geol.* 49, 785–94.

Smart, C. C. 1983: The hydrology of the Castleguard karst, Columbia Icefields, Alberta, Canada. *Arctic and Alpine Research* 15 (4), 471–86.

Smart, C. C. 1988a: Exceedance probability distributions of steady conduit flow in karst aquifers. *Hrdrological Processes* 2, 31–41.

Smart, C. C. 1988b: Artificial tracer techniques for the determination of the structure of conduit aquifers. *Groundwater* 26 (4), 445–53.

Smart, C. C. 1988c: Quantitative tracing of the Maligne Karst System, Alberta, Canada. *J. Hydrol.* 98, 185–204.

Smart, P. L. 1981: Variations of conduit flow velocities in the Longwood to Cheddar Rising System, Mendip Hills. *Proc. 8th Int. Congr. Speleo.,* Kentucky, 1, 333–5.

Smart, P. L. 1984: A review of the toxicity of twelve fluorescent dyes used for water tracing. *NSS Bull.* 46, 21–33.

Smart, P. L. and Hobbs, S. L. 1986: Characteristics of carbonate aquifers: a conceptual basis. In *Proceedings, Environmental Problems in Karst Terranes and their Solutions.* Bowling Green, KY: National Well Water Association, 1–14.

Smart, P. L. and Hodge, P. 1980: Determination of the character of the Longwood sinks to Cheddar resurgence conduit using an artificial pulse wave. *Trans. Brit. Cave Res. Assoc.* 7 (4), 208–11.

Smart, P. L. and Laidlaw, I. M. S. 1977: An evaluation of some fluorescent dyes for water tracing. *Water Resources Res.* 13, 15–23.

Smart, P. L. and Smith, D. I. 1976: Water tracing in tropical regions, the use of fluorometric techniques in Jamaica. *J. Hydrol.* 30, 179–95.

Smart, P. L., Atkinson, T. C., Laidlaw, I. M. S., Newson, M. D. and Trudgill, S. T. 1986: Comparison of the results of quantitative and non-quantitative tracer tests for determination of karst conduit networks: an example from the Traligill Basin, Scotland. *Earth Surface Processes and Landforms* 11, 249–61.

Stanton, W. I. and Smart, P. L. 1981: Repeated dye traces of underground streams in the Mendip Hills, Somerset. *Proc. Univ. Bristol Speleo. Soc.* 16 (1), 47–58.

Swinnerton, A. C. 1932: Origin of limestone caverns. *Bull. Geol. Soc. Am.* 43, 662–93.

Thrailkill, J. 1968a: Chemical and hydrological factors in the excavation of limestone caves. *Bull. Geol. Soc. Am.* 79, 19–46.

Trudgill, S. T. 1987: Soil water dye tracing, with special reference to the use of Rhodamine WT, Lissamine FF and Amino G Acid. *Hydrological Processes* 1, 149–70.

Vennard, J. K. and Street, R. L. 1976: *Elementary Fluid Mechanics*, 5th edn, S.I. version. New York: Wiley.

White, W. B. 1988: *Geomorphology and Hydrology of Karst Terrains.* New York: Oxford University Press.

Williams, P. W. 1977: Hydrology of the Waikoropupu Springs: a major tidal karst resurgence in northwest Nelson (New Zealand). *J. Hydrol.* 35, 73–92.

Williams, P. W. 1983: The role of the subcutaneous zone in karst hydrology. *J. Hydrol.* 61, 45–67.

Williams, P. W. 1984: Karst hydrology of the Takaka valley and the source of New Zealand's largest spring. In A. Burger and L. Dubertret (ed.) *Hydrogeology of Karstic Terrain: Case Histories.* International Contributions to Hydrogeology Series, Vol. 1. Series eds G. Castany, E. Groba and E. Romijn. Hanover: Heise.

Worthington, S. R. H. 1991: Karst Hydrogeology of the Canadian Rocky Mountains. Unpublished PhD thesis, McMaster University, 370 pp.

THREE

Processes of Cave Development

Introduction

Once water is circulating in the limestone, the roles of limestone solution, of rock architecture and of ceiling collapse in the evolution of a cave become dominant. The final plan and cross-sectional forms of an individual cave owe much to both the purity of the limestone and the network of fissures which dissect the rock. Although the gross morphology of a cave passage may be due to its hydrological setting, and variations in hydrology through time, at a local level the roles of lithology and structure become important in guiding solutional processes. Following a brief treatment of limestone lithology and structure, the solution process in the dominant cave-forming rock – limestone – is outlined. The solution of dolomite, gypsum and sandstone is also addressed. The translation of water circulation in the rock into the cave passages we can observe is covered in detail, including the role of warm mineralized water. Some examples of cave passage shapes in different limestone settings are also given. Geological history plays a key role and may override other factors. In this chapter examples are provided from the largest and deepest caves known, in Sarawak (Mulu) and France (Réseau Jean-Bernard).

Karst Rocks

Limestone

Approximately 17 per cent of the earth's land surface is made up of carbonate rocks, with the major carbonate regions being in Europe, eastern North America, and East and South-East Asia (see figure 3.1), and the remnants of Gondwanaland being relatively depauperate in carbonates because of their extreme age. Because of overlying deposits, unsuitable climate or low relief, only 7 to 10 per cent of that area displays karst landforms (Ford and Williams

1989, 6). Despite this modest extent, about 25 per cent of the world's population is dependent on karst groundwater. This dependence is increasing in the countries of Asia where population growth is still rapid.

Limestones are generally regarded as being composed of more than 50 per cent calcite or calcium carbonate, with pure, cave-forming limestones generally having more than 90 per cent calcite (see figure 3.2). Dolomites, the calcium magnesium carbonate rocks, follow a similar classification. Caves are less numerous in dolomite on account of its lower solubility under ambient conditions. Below 50 per cent calcite or dolomite it is unlikely that caves will form at all, though isolated examples form along tectonic fissures in such rocks. Limestones range in age from Precambrian to Holocene, with modes in the Ordovician, Carboniferous, Cretaceous and mid-Tertiary (see plate 3.1). Limestones can form in both marine and freshwater environments, but most have formed in shallow warm to tropical seas. Limestones can also be divided into facies depending on the precise environment of their formation: for coral reefs, these are shown in figure 3.4.

One of the most widely used classifications of limestone comes from the work of Folk (1962). This classification concentrates on

Figure 3.1
Global distribution of carbonate rocks. From Ford and Williams (1989).

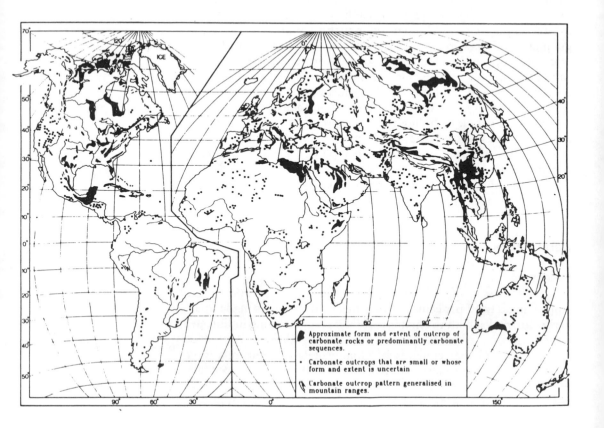

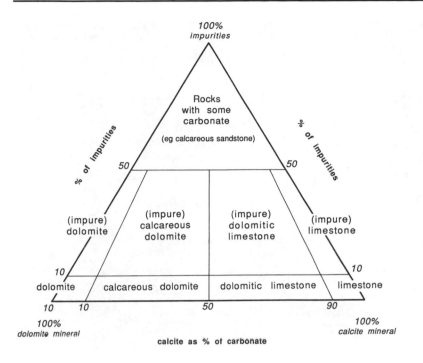

Figure 3.2
Classification of carbonate rocks by relative proportions of calcite, dolomite and impurities. After Leighton and Pendexter (1962).

the nature of the transported grains (allochems) and the calcite cements formed *in situ* (orthochems). Other components may include terrigenous material such as sands and clays and heavy minerals such as zircon and rutile. Allochems may be subdivided as follows:

1 intraclasts: reworked fragments of older carbonates
2 oolites: rounded forms with a concentric structure
3 pellets: usually of faecal material
4 fossils: corals and shells, sometimes vertebrates.

Orthochem cements may be either micrite (see plate 3.2), a fine opaque ooze of microcrystalline carbonate, or sparite, semi-translucent rhombic crystals of calcite. The detail is given in table 3.1 and figure 3.3. Depending on the proportions of allochems, sparry cement or micritic cement, the limestone may fall into zones on the ternary diagram. Its only disadvantage is that it does not take account of diagenesis (chemical alteration after deposition) and is therefore difficult to apply in some instances. A massive concretion of fossil material, such as coral colonies or algal mats in growth position or as reef talus, may be called a biolithite (see plate 3.3). Another simple classification is based on the median grain size (see table 3.2). This classification is useful for the younger dune limestones or calcarenites which contain cemented allogenic clasts of sand or silt size, including fossil shell or bone material.

Plate 3.1
Thinly bedded Tertiary
limestone with stylolites at
Punakaiki, New Zealand.

The genetic relationships between different limestone facies may be apparent in gorge sections and sometimes in caves. The classic example is the limestone of the Napier Range in the Kimberley of Western Australia, a Devonian reef complex formed along a shoreline shaped in Precambrian rocks. This reef complex included fine textured inter-reef facies deposited in deep water, the tumbled blocks of the fore-reef slope or reef talus, the *in situ* bioherms of the growing reef rim and the muddier, flat-bedded back-reef facies of the lagoon (see figure 3.4). Following uplift, Tertiary planation and incision, a cross-section through this reef complex is exposed in the gorges of the Lennard and Fitzroy rivers. The karst morphology is affected by the facies type, with rugged 'giant grikeland' terrain formed on the fore-reef and reef-rim facies contrasted with subdued relief on the muddier back-reef facies. Fissure caves and network mazes are numerous in the fore-reef facies, while caves in the lagoon facies are scarce and associated with fluvial incutting. Reef talus and lagoon facies limestones are well exposed in the walls of Rope Ladder Cave, near Townsville, Queensland; the jumbled blocks of fore-reef facies have been bevelled by solution under phreatic conditions (see plate 3.4).

Limestones are very susceptible to diagenesis because of their solubility at ambient temperatures and pressures. The circulation of

Table 3.1 Classification of limestones after Folk (1962)

| Allochems | Orthochems | |
	Micrite	Sparite
None	Micrite	Sparite
Intraclasts	Intramicrite	Intrasparite
Oolites	Oomicrite	Oosparite
Pellets	Pelmicrite	Pelsparite
Fossils	Biomicrite	Biosparite

Plate 3.2
Massive fossiliferous micritic limestone with brachiopods and breccias of Silurian age at Yarrangobilly, New South Wales. The scale bar is 10 cm long.

acidified water through limestones creates endless opportunities for solution and recrystallization in both small and large voids. Virtually all ancient limestones have experienced at least one such phase of diagenesis, which is both rapid and extensive. Following exposure of the limestone by uplift or sea-level change, circulating freshwaters dissolve the carbonate, especially the aragonite. This is succeeded by sparry calcite cements, often slowly by isomorphous replacement so that the 'ghosts' of aragonite fossils are preserved. In contrast, rapid replacement removes the fossil traces, substituting sparry calcite throughout. Voids along bedding planes may be infilled by cements and sediments, including vein mineral deposits such as those of the Derbyshire limestones. Under strong conditions of evaporation, calcrete nodules and crusts, gypsum and pisolites may be deposited in the upper strata of the limestone.

Shallow water limestones may often contain interbeds of soluble minerals such as halite or gypsum, which upon diagenesis are removed, leading to partial collapse of the limestone. Minor fissures

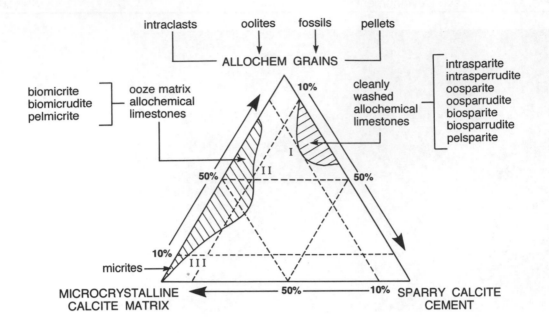

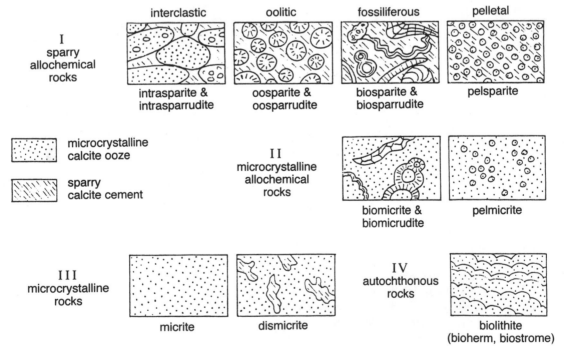

Figure 3.3
Major types of limestone. According to Folk (1959).

Plate 3.3
*Gently dipping Miocene
limestone in the Muller
Range, Papua New
Guinea, has pure limestone
units 30 to 50 m thick
separated by muddy facies
which act as aquicludes.*

Table 3.2 Classification based on the median grain size

	Median grain size	
calcrudite	>2 mm	e.g., coral or shell fragments
calcarenite	0.02–2 mm	e.g., sand-sized material
calcilutite	<0.02 mm	e.g., fine-grained silts and clays

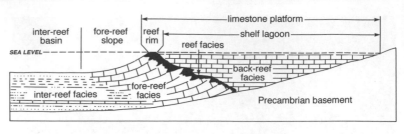

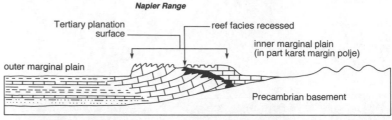

Figure 3.4
Facies types for carbonate reefs, with an example from the Napier Range, west Kimberleys, Australia. After Playford and Lowry (1966).

Plate 3.4
Reef talus limestone of Devonian age exposed in the walls of a formerly phreatic cave at Fanning River, north Queensland.

may also open in the rock because of uplift. These voids may be subsequently infilled by coarse material, including limestone fragments, creating breccias. The cavities surrounding the clasts are often infilled by sparry calcite or other minerals, including phosphates and limonites. Coarse breccias can also form as reef talus (fore-reef) limestones with micritic cavity infillings. High temperature and pressure may also metamorphose limestones to marble, a mosaic of large clear calcite grains that interlock. However, some limestones may also recrystallize by solution without high temperature and pressure to give coarse calcirudites in which the grains do not interlock.

Nodules or thin sheets of chert or flint will often form in lime-rich sediments by the dissolution of quartz or silica from sponges, diatoms and radiolarians. Nodules can be up to 1 m in diameter, especially in the chalky Wilson's Bluff limestone of the Nullarbor Plain. Chert bands are usually less than 50 cm thick and, being relatively brittle, are fractured, allowing some water movement. Stylolites are pressure solution seams formed under conditions of deep burial of the rock. They are often darker than the host lime-stone because of the concentration of insoluble minerals and or-ganic matter along them. Stylolites are usually a few millimetres to a few centimetres thick, and generally individual forms are only a few tens of centimetres long.

Dolomite

The formation of dolomite is a complex subject which is still not fully understood. Dolomite forms by the replacement of existing calcite and aragonite. There are four principal modes of formation: reflux, mixing, burial and hydrothermal. Seawater concentrated by evaporation in lagoons may reflux through limy sediments, ex-changing ions with them. This process is enhanced where there is a high ratio of magnesium to calcium and abundant carbonate ions in the crystal lattice, breaking the bond of the strongly hydrated magnesium ions. This process may occur very early in diagenesis. Under mixing conditions of fresh- and seawater, where calcite is soluble but dolomite is not, the necessary magnesium : calcium ratio is met and dolomite precipitates. Hanshaw and Back (1979) pro-pose that this is occurring in the mixing zone deep in Florida karst groundwater. Deeply buried carbonates may be dolomitized by the expulsion under pressure of fluids, but the extent of this process is disputed. Hydrothermal dolomites tend to be located along fracture zones where warm basin fluids are circulating.

Karst development on dolomites is generally restricted, though significant caves are found in that host rock. The caves of western Transvaal, South Africa, are formed in Proterozoic dolomites and cherts and include Apocalypse Pothole, which is 12 km long. In the American states of Wisconsin and Minnesota there are many caves formed in the Prairie du Chien dolomite, while the Barkly Tableland of Queensland and the Northern Territory, Australia, contains extensive horizontal caves which intersect the regional water table.

Sandstone

Sandstones are essentially granular rocks comprised of silica grains cemented by calcite, clays, iron oxides or silica. The mineral com-position of the grains in the sandstone can provide much informa-tion about the source areas – granitic, metamorphic, volcanic or sedimentary. Immature sandstones contain readily weathered minerals such as feldspars and ferromagnesian minerals, and imply deposition close to their source. In contrast sandstones dominated by quartz imply maturity and/or long-distance transport. Four common types of sandstone are:

- quartz sandstone (dominated by well-sorted and rounded quartz with little or no feldspar, mica or fine clastic material)
- arkose (dominated by angular quartz grains, but with more than 25 per cent feldspar, bonded by calcareous cements, clays or iron oxides)
- greywacke (poorly sorted immature sandstones in a finer matrix of dark clays, silt, micas and chlorite)
- subgreywacke (sandstones in which better-sorted and rounded quartz grains sit in a matrix of not more than 15 per cent fine cement in which feldspar is scarce).

Large caves may be found in quartz sandstones and arkoses but rarely in greywackes. Sandstones may be very fine grained or quite coarse, containing a wide range of grain sizes depending on the energy and persistence of transport. The sedimentary structures, usually well preserved, can give quite precise information of the environment of deposition – alluvial bars or fans, deltaic plains, turbidity currents, shallow or deep water. The primary porosity of sandstones is a function of the mean grain size of the quartz grains and the nature of the cement. Sandstones are frequently highly fractured and the joint networks in them often resemble those of limestone massifs. Silica is just soluble in meteoric waters and will dissolve, slowly; however, often the cement is more soluble and will rapidly dissolve, allowing granular disintegration. Once subject to metamorphism they may be wholly or partially converted to finer grained quartzites, in which the grains interlock. These hard rocks are less soluble, but significant caves form in them, primarily in Venezuela and Arnhem Land, Australia.

Processes of Dissolution of Karst Rocks

The solution of limestone in meteoric waters

The slow dissolution of limestone is a superficially simple process in which two minerals, calcite and dolomite, are dissociated in acidified water. Although organic and mineral acids may be very important in certain circumstances, under most contemporary conditions the dissolution of limestone is dominated by carbonic acid formed from dissolved carbon dioxide. The sources of this carbon dioxide will be considered later. There is a sequence of reactions, summarized in table 3.3. The last equation, (3.5), disturbs the equilibrium of (3.4) by the removal of CO_3^{2-} so that more carbonate must dissociate to restore the balance. In addition the association of H^+ + CO_3^{2-} disturbs the equilibrium in (3.3), promoting further dissociation. This in turn disturbs the equilibrium in (3.2) and ultimately (3.1), causing more carbon dioxide to dissolve in water.

These processes continue until the forward and reverse reaction rates are equal, at which point the system is in equilibrium and the solution is saturated with respect to calcite. Any acid (be it carbonic, organic or inorganic) will add hydrogen ions to the system and will displace (3.3) and (3.5) in a forward direction, reducing the concentration of CO_3^{2-} and thus permitting more dissolution of calcite.

If all these equations are combined we get the commonly quoted dissolution equation for calcite:

$$CaCo_3 + H_2CO_3 \Longleftrightarrow Ca^{2+} + 2HCO_3^{-} \qquad (3.6)$$

For dolomites, the presence of magnesium ions (Mg^{2+}) in equation (3.3) complicates the dissolution process, which is generally written as:

$$CaMg(CO_3)_2 + 2H_2CO_3 \Longleftrightarrow Ca^{2+} + Mg^{2+} + 4HCO_3^{-} \quad (3.7)$$

There are no thresholds in these reactions – they will occur in both static and moving water – in contrast to the mechanical erosion by water of other rocks, where there is a threshold for transport of weathered material. But other factors such as temperature and gas concentration may have an effect on limestone solution.

The solubility of carbon dioxide gas in water decreases as temperature increases, in line with Henry's law, at a rate of about 1.3 per cent per degree Celsius. Although CO_2 is more soluble in cold water, there is much less soil CO_2 available in arctic regions and the colder water makes the reaction proceed more slowly. The work of Ford and Drake (1982) shows that the gas concentration of CO_2 and the amount of water moving past the rock–water interface are more important than the absolute concentration of dissolved CO_2. Thus limestone solution rates would tend to be greater in a tropical

Table 3.3   Simplified processes of solution of carbonates

Process equation	Kinetics	Description
$CO_2 \Longleftrightarrow CO_2$ air dissol	slow	diffusion of CO_2 into water (3.1)
$CO_2 + H_2O \Longleftrightarrow H_2CO_3$ dissol	slow	hydration of dissolved carbon dioxide to form carbonic acid (3.2)
$H_2CO_3 \Longleftrightarrow H^+ + HCO_3^-$	fast	dissociation of carbonic acid into hydrogen and hydrogen carbonate ions (3.3)
$CaCO_3 \Longleftrightarrow Ca^{2+} + CO_3^{2-}$	slow	dissociation of calcite crystal lattice to ions (3.4)
$H^+ + CO_3^{2-} \Longleftrightarrow HCO_3^-$	fast	association of carbonate ions with hydrogen ions to form a hydrogen carbonate (3.5)

climate where there is a larger soil CO_2 concentration on account of bacterial activity, and the rainfall is much higher.

We can consider limestone solution to occur under two contrasted situations: an *open system*, where carbon dioxide gas, percolation water and calcite-rich rock are continuously in contact, and a *closed system*, where carbon dioxide gas and water equilibrate but the supply of gas is cut off before the water contacts the rock. From equations (3.1) and (3.2) there is no replacement of CO_2 under closed conditions, and thus the total amount of calcite that can be dissociated is much less. This is shown in figure 3.5. Note also the effect of temperature on the equilibrium solubility of calcite in an open system.

Thus any body of percolation water contacting limestone will gradually reach saturation, beyond which no more dissociation of the rock will occur unless conditions change. However, one way in which this may occur is if two bodies of saturated water mix – as they may do at tributary junctions in a cave or below the water table in the phreatic zone. The process, first described by Bögli (1964), is termed mixing corrosion. If each body of water is saturated with respect to calcite at different partial pressures of CO_2 (PCO$_2$), then by mixing they will produce a new solution which is undersaturated (see figure 3.6). In extreme cases an extra 20 per cent more calcite could be dissolved, but the norm is more like 1 to 2 per cent. When vadose seepage water at high

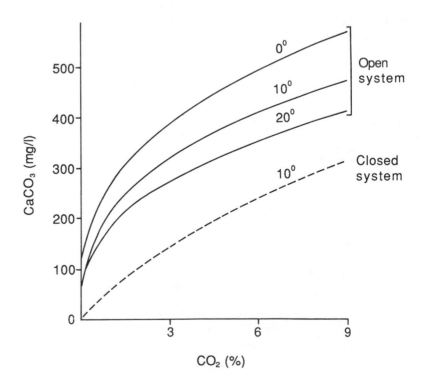

Figure 3.5
Equilibrium solubility of calcium carbonate in contact with air containing carbon dioxide for open and closed systems at varying temperatures. After Picknett et al. (1976).

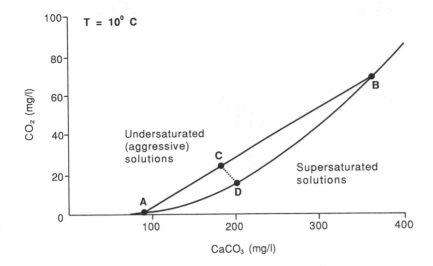

Figure 3.6
*The 'mixing corrosion'
principle: solubility of
calcite with respect to
carbon dioxide in solution.
The curve ADB shows the
solubility of calcium
carbonate with respect to
dissolved carbon dioxide.
Mixing of two saturated
solutions A and B
produces a solution C
which is undersaturated.
Solution C will then evolve
to saturation along the line
CD. Based on Gunn
(1986).*

PCO_2 meets a vadose stream at lower PCO_2, then a substantial increase in limestone solution is possible and may be expressed as dilation of the cave passage. At great depth, percolation water at high PCO_2 may encounter saturated water moving slowly. This form of mixing corrosion may be important for conduit enlargement in the early stages of cave development (Dreybrodt 1981).

Soil and vegetation in the limestone solution process

Most textbook discussions of the limestone solution process tend to concentrate on chemical kinetics and on hydrology, especially those parts of the process occurring within active cave conduits. A notable exception is Trudgill's (1985) text on limestone geomorphology. Yet the veneer of soil on most limestones has a central role in karst processes, through its control in water infiltration and storage, by acting as a CO_2 generator and through the role of soil buffering capacity in the solution process. The roles of soil and vegetation can be outlined using figure 3.7. Healthy perennial vegetation acts to intercept rainfall and its dissolved gases, principally carbon dioxide (CO_2). Atmospheric carbon dioxide is currently 0.034 per cent of the total volume and is increasing at approximately 5 per cent of its concentration per annum. To this can be added other aerosols such as sulphuric acid from industrial sources and often oxides of nitrogen formed in thunderstorms. The patchy cover of vegetation intercepts rain and protects the soil surface from rain and wind erosion. At the ground surface the infiltration of rainwater is aided by leaf litter and hollows which allow most water to percolate into the soil; the excess runs off to surface channels. The complex of decaying vegetation, soil fauna

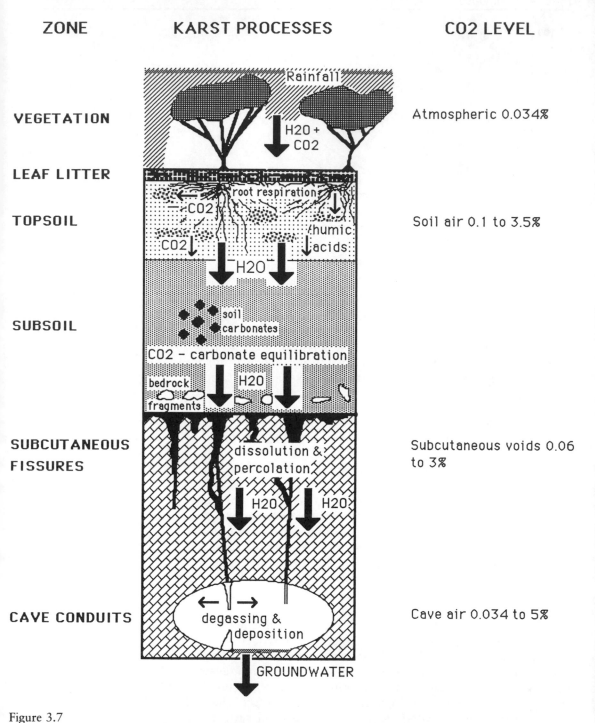

Figure 3.7
The cascade of carbon dioxide through the vegetation, soil and subcutaneous zones of karst. Not all zones may be present in any single karst, and only the range of recorded carbon dioxide concentrations in temperate ecosystems is given.

(including bacteria) and fungi in leaf litter provides a source of carbon dioxide and additional organic compounds such as humic and fulvic acids. In the root zone the respiration of up to 25 per cent of the carbon dioxide taken up by plants occurs. This released gas may dissolve in percolating rainwater. Bacteria in topsoil also release copious quantities of the gas through their metabolism, and are the major producers. In acidic soils bacteria are disadvantaged: decomposition of organic matter is achieved by fungi which are much slower yet also yield carbon dioxide. These processes result in soil CO_2 concentrations of between 0.1 and 3.5 per cent in temperate regions, and up to 10 per cent in the tropics. Much of this gas dissolves in soil water to give a weak acid which is carried down by gravity to the subsoil along with humic and other organic acids.

In the subsoil aeration is much less and bacterial action is reduced owing to this and lower organic matter content. The important process of carbon dioxide–calcium carbonate equilibration takes place in this zone. Carbon dioxide gas dissolves in water to give the weak carbonic acid, which dissociates to hydrogen and hydrogen carbonate ions. An increase in CO_2 concentration in the soil will allow more hydrogen ions to be released into drainage water, acidifying the soil (lower soil pH). This greatly enhances the ability of the percolating water to dissolve calcium carbonate. Some or all of the weak acid may be neutralized by exchangeable calcium ions released from clays, from carbonate concretions or from bedrock fragments. Thus bedrock erosion is much reduced where the soil has a high exchangeable calcium content released from leaf litter, or where it contains a high proportion of bedrock fragments. Changes in soil pH may be reduced by the reserve, or buffering, capacity of hydrogen ions derived from organic acids released from the decomposition of clay-humus particles and organic matter. The chemical balance of the karst solution process is thus very dependent on organic influences and especially the maintenance of vegetation, leaf litter and a productive topsoil. Where native, calcium-loving vegetation is replaced by plants adapted to more acidic soils, then the soil may become acidified to bedrock and accelerated erosion of limestone occurs. This has been the case in Yorkshire, where more frequent burning has encouraged the growth of heather at the expense of the natural grassland and forest vegetation: there are now deep runnels formed in bedrock down which soil can be lost (Trudgill 1976). A similar process has been observed in New Guinea, where slash and burn cultivation on limestone is accompanied by greatly accelerated soil loss (Gillieson et al. 1986).

The zoning of solution in the unsaturated zone

Below the soil zone in karst, there is a high porosity zone of weathered rock defined by Williams (1983) as the subcutaneous zone (see plate 3.5). This zone has also been called the epikarst

(Mangin 1975; Friederich and Smart 1981; Bakalowicz 1981). Coupled with the soil, this zone stores large quantities of water and is the site of a high level of solutional erosion. Gunn (1981) measured the solutional load of the components of the epikarst water at Waitomo, New Zealand, and his study provides perhaps the most comprehensive set of data yet obtained on solutional processes in the upper skin of the karst. In this extensive polygonal karst subcutaneous flow, shaft flow and some vadose flow had hardness values in the same range as cave streams (see table 3.4).

Plate 3.5
Enlarged joints in the epikarst are exposed in marble quarry walls at Wombeyan, New South Wales.

Allogenic water flowing onto karst is highly capable of both chemical and mechanical erosion. Both allogenic and autogenic components of solutional denudation need to be separated if sense is to be made of rates of landscape development. This requires at least a partial solutional budget for the different components of the karst drainage system (see figure 3.8). In the Riwaka basin of New Zealand, karst rocks occupy just less than 50 per cent of the basin, and autogenic solution is about $79\,mm\,ka^{-1}$; from solute concentrations, much of this occurs in the superficial zone. The large allogenic component increases total basin solution by about 20 per cent.

Both Gunn (1986) and Smith and Atkinson (1976) concluded that 15 to 50 per cent of the total solutional erosion occurred in the deeper endokarst zone, which contains the cave conduits. However, a consideration of karst porosity by Worthington (1991) indicated that the effective void volume of the endokarst was around 1 to 1.4 per cent only, while at the top of the epikarst it increased rapidly to 100 per cent at the surface. Thus only about 1 per cent of the endokarst has been removed by solution processes, not the 15 to 50 per cent cited earlier. Cave conduits formed by the solutional attack of circulating groundwater therefore occupy only a very small percentage of the total voids in the karst.

Limestone solution in seawater

In the marine environment limestone dissolution is greatly enhanced owing to the common ion effect. Addition of large quantities of foreign ions such as Na^+, K^+ and Cl^- to a bicarbonate-rich

Table 3.4   Total hardness measurements at Waitomo, New Zealand (after Gunn 1981)

Hydrologic component	Number of samples	Mean water hardness $mg\,L^{-1}$	Standard deviation $mg\,L^{-1}$
Rainfall	26	5	4
Overland flow	6	21	20
Soil water	109	64	10
Throughflow	154	51	14
Subcutaneous flow	139	122	18
Shaft flow	58	122	18
Vadose flow (Mangapohue catchment)	84	130	15
Vadose flow (Glenfield catchment)	99	40	5
Vadose seepage	1714	96	19
Cave stream: Mangapohue	59	125	8
Cave stream: SP2	33	126	7
Cave stream: Glenfield	59	124	9

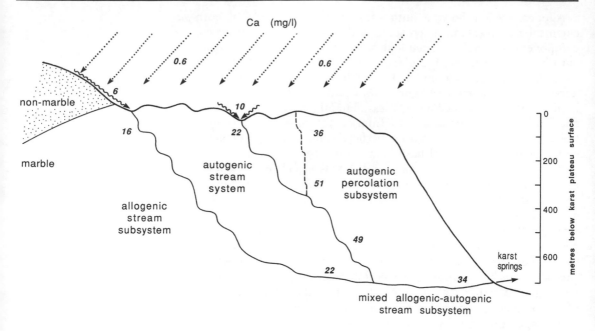

Figure 3.8
*Dissolved calcium
concentrations in the
Riwaka karst, New
Zealand. From Williams
and Dowling (1979).*

water decreases the activity of Ca^{++} and Mg^{++} ions, thus increasing
the rate of dissolution of the solid phase carbonate or dolomite. The
presence of 0 to 10 per cent Mg^{++} reduces Ca^{++} solubility, while
greater than 10 per cent Mg^{++} increases Ca^{++} solubility (Picknett et
al. 1976). In seawater the addition of large concentrations of sodium chloride greatly enhances limestone solution. In warm water
with high partial pressures of carbon dioxide, calcite solubility can
be boosted to approximately 1000 mg L^{-1} (Plummer 1975). Where
limestone coasts occur, a mixing zone between fresh and marine
groundwaters exists. For the Yucatan peninsula of Mexico, Back
et al. (1984) recorded the long-distance flow (100 km) of
saturated karst groundwater with calcite concentrations of around
250 mg L^{-1}; in the final kilometre of flow, an additional 120 mg L^{-1}
is added owing to mixing with a seawater lens. This coastal mixing
zone may be a site for intensive solution cavity formation, producing a narrow zone of maze caves which may invade pre-existing
cavities in raised reefal limestone.

Careful examination of contemporary coral reefs shows that
even very young limestones, actively undergoing diagenesis, have
extensive rounded voids which are elongated along proto-bedding
planes. In most cases the corals are still clearly visible in either
biohermal structures or in reef talus. These voids are interconnected
to form networks through which seawater, and later freshwater,
may circulate. These structures commonly survive uplift and may
guide meteoric waters to produce mazes of solution cavities, some
of which may be large enough for human entry. Further diagenesis
may radically alter the original fossils and depositional structures,
but the cavities tend to persist and enlarge.

The form of the freshwater lens on limestone islands is an important determinant of the types of cave that form there (Nunn 1994).
In the previous chapter the role of the Ghyben–Herzberg lens
was emphasized in promoting limestone solution in a fresh–
marinewater mixing zone. Below the water table phreatic passages
form with a series of chambers linked by short passages, often
complex in shape. The walls of these passages are frequently scalloped (Ollier 1975). Recently caves of this type have been entered
and mapped by divers (Palmer 1989). At the water table epiphreatic
caves form in a fluctuating zone, and display characteristics of both
phreatic and vadose caves. These may often connect with sea caves.
Finally, vadose caves fed by rainwater may form, draining to lower
phreatic systems. Thus the caves in uplifted coral terraces combine
primary reefal structures with solution features on account of rainwater inflow or inherited from previous phreatic conditions. Often
stream invasion, lake or shoreline incuts may further modify the
form of the cave. Reactivation of entrenchment, wall collapse or
infilling may follow tectonic uplift or sea-level change. The occurrence of phreatic caves well above the water table is evidence of
emergence, possibly uplift, while submerged caves may suggest
island subsidence. Thus in Tuvalu, Gibbons and Clunie (1986)
described a cave 46 m below present sea level with evidence of
human occupation. These coastal caves are among the simplest in
the spectrum of cave morphologies, yet from the preceding discussion their origins are diverse and complicated and they may tell us
much about relative movements of land and sea. The more familiar
caves in the older limestones of major landmasses are an order of
magnitude more complex to unravel.

Solution of evaporites

All the evaporite rocks – gypsum ($CaSO_4.2H_2O$), anhydrite
($CaSO_4$) and halite ($NaCl$) – are soluble in pure water, producing
little or no residue on dissociation, and they may be regarded as
karst rocks. Evaporite karst will survive under arid or semi-arid
conditions only because of this high solubility. In addition, the large
volume expansion of anhydrite on hydration to gypsum acts to
close developing karst conduits (Jakucs 1977). Thus evaporite
karsts are rare on a global scale, but within them true karst
landforms of karren, dolines, stream caves and springs may form
(Wigley et al. 1973). Of more significance to the world of karst is
the role of both gypsum and halite in the process of exsudation,
where expansion on crystallization breaks other karst rocks and
enlarges cave passages. The deposition of both gypsum and halite
speleothems is treated in Chapter 4.

Solution of silicates in meteoric waters

Silicate rocks such as sandstones and quartzites are reasonably
soluble in natural waters, with the solubility increasing markedly in

alkaline conditions (Young and Young 1992, 62). Although under normal conditions of acidified drainage waters quartz is less soluble (about $30\,mg\,L^{-1}$ or $0.5\,mmol\,L^{-1}$ at $25°C$), over a long time significant solution will occur. The weathering of silicate rocks involves both the dissociation of silica to silicic acid and the weathering of constituent minerals and cements to produce solutes and clays.

The dissolution of quartz can be written as follows (Young and Young 1992, 61):

$$SiO_2 + 2H_2O \Longleftrightarrow H_4SiO_4 \qquad (3.8)$$

The weathering of a component mineral such as a feldspar to gibbsite or kaolinite can be simplified to:

$$NaAlSi_3O_8 + H^+ + 7H_2O \Rightarrow Al(OH)_3 + Na^+ + 3H_4SiO_4 \quad (3.9)$$

$$NaAlSi_3O_8 + H^+ + 7H_2O \Rightarrow \frac{1}{2}Al_2Si_2O_5(OH)_4 + Na^+ + 2H_4SiO_4$$
$$(3.10)$$

In both cases silica is mobilized in solution as weak monosilicic acid. This may precipitate as amorphous silica or opal-A, or combine with other compounds to form more complex clays.

$$2Al(OH)_2^{\;+} + 2SiO_2 + H_2O \Rightarrow Al_2Si_2O_5(OH)_4 + H^+ + H_2O$$
$$(3.11)$$

Amorphous silica is more soluble than the quartz: its solubility ranges from 60 to $80\,mg\,L^{-1}$ ($1–1.3\,mmol\,L^{-1}$) at $0°C$ to 100 to $140\,mg\,L^{-1}$ ($1.7–2.3\,mmol\,L^{-1}$) at $25°C$ over a wide range of pH. In general silicate minerals weather more readily than amorphous silica, which itself is more soluble than quartz. Thus the mineral composition and grain size of the silicate rock affects its weathering rate quite significantly.

Thus in the weathering of silicate rocks there is far more insoluble material produced, which may infill developing caves and dolines. These residues block the interstices between grains and infill small joints, reducing the permeability of the silicate karst. Surface solution features such as solution pans and runnels may be seen on many sandstones, especially in the tropics. The 'beehive' terrain developed in tropical sandstones such as the Bungle Bungles (Purnululu National Park) in Western Australia in some ways mimics polygonal karst (Young 1989), but subterranean drainage is restricted or absent. Large stream caves seem to be restricted to situations where prominent joints or faults combine with steep hydraulic gradients, such as near plateau edges. One example of this type is the Sima Aonde in Venezuela, where very large dolines with solution pans (cinegas) combine with large stream caves showing strong joint control (see figure 3.9). The quartzite caves of

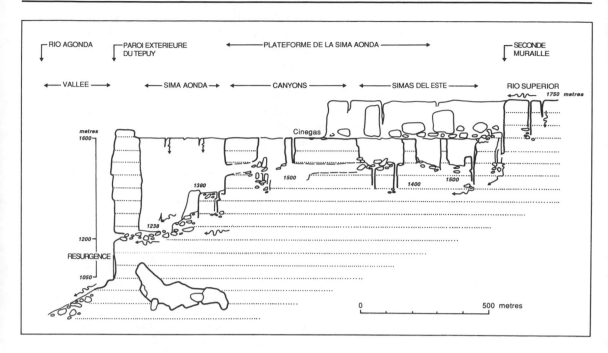

Figure 3.9
Underground circulation of the sandstone karst of the Sima Aonde, Venezuela. Simas are deep pits; cinegas are a type of solution pan. From Courbon et al. (1989).

Zimbabwe, including the 305 m deep Mawengc Mwena (Truluck 1994), extend the realm of deep caves in sandstone. Thus Jennings (1983) concludes that this particular case is true karst, and is deserving of further study. It would be instructive to compare karst forms and solution processes in sandstone and limestone existing in a similar tectonic setting and under a similar climate. Care must be taken to differentiate between true silicate karst and subjacent karst where limestone underlies the sandstone, and both solution and collapse in underlying cavities occur – for example, Big Hole, New South Wales (Jennings 1967).

Rock Control and Cave Morphology

Role of lithology

Quite often caves are guided by changes in lithology, with passages developing along or close to the junction of pure and impure limestones, limestones and underlying shales, or limestones and igneous rocks such as granite. There may be a stratum or horizon in which passage development occurs preferentially, even at the proto-cave stage; this is what Lowe (1992) calls the inception horizon. In the Forest of Dean (UK) the major cave inception horizons are associated with interbedded sandstones and unconformities in the Carboniferous limestone (Lowe 1992). In McBride's Cave (Cumberland Plateau, Tennessee) flat-lying chert beds serve as aquicludes and result in a stepped long-profile with long, low epiphreatic tubes linked by waterfall shafts located at major joint intersections. Such an impure or impermeable bed need not be continuous: for example, discontinuous cherty layers in the Tertiary limestones of New Guinea act as aquicludes, and cave passages form just above them, as in Atea Kananda (Muller Range; Gillieson 1985). Strong's Cave (Western Australia) is formed at the junction of the Tertiary dune limestones and the underlying Proterozoic granite, which is partially exposed in the floor where speleothems are absent.

It is now very clear that there is an important role for impurities in limestone in guiding water flow in the karst. Insoluble beds in limestone act to confine water flow, especially in the vertical plane, and result in perched cave passages of varying sizes. This may occur at the earliest stages of cave formation – as the inception horizons of Lowe (1989, 1992) – or much later in the cave's history, when cave passage incision owing to external valley downcutting is retarded by resistant or insoluble strata, producing stepped passage profiles or cave waterfalls, such as that of Barber's Cave, Cooleman Plain, Australia, or Gaping Ghyll, Yorkshire (Jennings 1968). Impure muddy limestones occur every 20 to 30 m in the stratigraphy of the limestones of the Muller Range, Papua New Guinea, and act as aquicludes to direct cave passages (see plate 3.6) to many levels over a vertical range of 525 m in the Mamo syncline (Francis et al. 1980). Stylolitic bands owing to pressure solution of impurities are widespread in the Ordovician limestones of Tasmania, and form waterfalls in Herbert's Pot and Khazad-Dum caves. Where the limestone strata have been tilted, stylolitic bands can force rapidly descending and ascending phreatic tubes which create inverted siphons, such as those in Murray Cave and River Cave, Cooleman Plain, or in Peak Cavern, Derbyshire.

Although the development of caves is determined principally by hydraulic gradient and catchment size, much of the passage morphology and network architecture is determined by lithology and structure. Thus karst systems tend to be more random in pattern in flat-lying porous limestones than in steeply dipping crystalline limestones. This randomness has been investigated by Laverty (1987), who concludes that cave lengths exhibit fractal behaviour, with the fractal dimension in the range 1 to 1.5 over the compass of resolutions from 1 to 100 m for surveyed caves. The implication of this for speleology is that such relationships developed for individual caves may be applicable to a region, including unexplored or unenterable caves. Curl (1986) has investigated this possibility for the Appalachian karst and has developed regional area and volume equations of use to hydrology and biology.

Role of joints, fractures and faults

Water circulation within a karst is greatly enhanced when a network of joints is present. Joints are simple fractures in the rock with

no, or minimal, movement of the strata. They are usually caused by tensional and shear forces, and may be the result of diagenesis, uplift, folding or valley-side unloading following erosion. Most joints run at right angles to bedding, but they may be inclined (see plate 3.7) or even sinuous. In plan they may intersect to give a rectangular or rhomboidal network, with cross joints penetrating only a few beds and master joints extending through many beds. Master joints may be several hundred metres long and caves may form preferentially along them; for example, Lancaster Hole, York-shire, is partly guided by vertical joints.

Small joints may be tight and relatively impermeable, or may be filled with sediments or calcite. Large joints start as angular, irregu-lar cavities but become rounded by solution with time. Cave devel-opment is enhanced where the joint spacing is wide to very wide (100 to 300 m) because of the concentration of flowing water. Most caves will show some passages with joint network guidancc (see figure 3.10). In extreme examples, the joint network dominates to give a maze cave, such as Wind Cave, South Dakota, or Optimisticheskaja, Ukraine.

Joint sets are not static in time; any rock mass may bear the imprint of many phases of joint development, as a result of tectonism or erosional unloading. Large, deep tensional joints, tens of metres deep, often form at the edges of plateaux or valleys owing

Plate 3.6
An active phreatic tube in the Rinse Cycle, Mamo Kananda Cave, Papua New Guinea. This passage floods daily following afternoon storms.

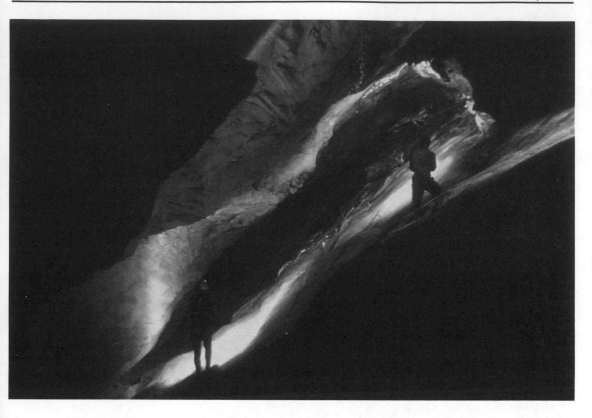

Plate 3.7
*A rift passage developed in
steeply dipping folded
marble in Greftkjell Cave,
Norway.*

to pressure release following scarp retreat. These may fortuitously
intersect active or fossil phreatic passages in the limestone mass.
Often the sequence of this joint development can be ascertained by
careful examination of cave passage intersections.

Although in a single passage the role of bedding and joints may
be reflected in the plan and cross-section of a cave, we need to
consider major geologic structures when we attempt to explain the
complex patterns of caves in length and depth. The increased
circulation of water in and around faults is an obvious factor in
karst system development. Faults are fractures with some displacement of the rock strata evident; a fault gangue of ground-up rock
may be present to clarify the sense of movement (see plate 3.8).
Such zones may act as inception horizons for proto-cave development, and may continue to dominate the array of conduits. Several
of the world's deepest known shafts (see figure 3.11) are located in
fault zones; the process of solutional enlargement is aided by the
periodic collapse of weaker fault gangues and the increased vertical
permeability. Thus the deep shafts of Mavro Skiadi, Provatina and
Epos in Greece are all of this type. In the last two frost shattering
also plays an important role in passage enlargement. Individual
cave chambers may be guided by faults – for example, the main

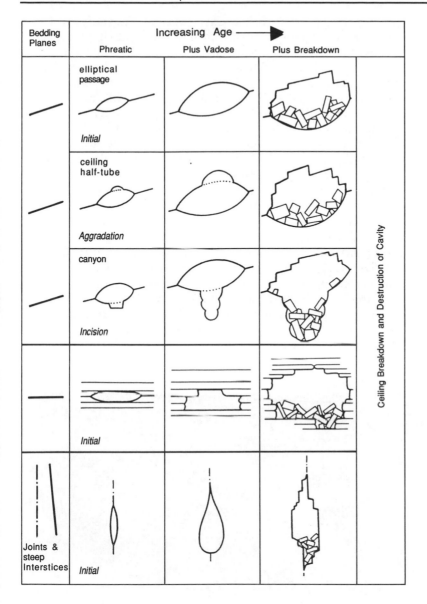

Figure 3.10
Variation in passage shapes in relation to bedding and joint orientation with increasing age of phreatic and vadose cave passages. From Bögli (1980).

chamber of Gaping Ghyll, Yorkshire; West Driefontein Cave, South Africa; P8 Cave, Peak District; and Marakoopa Cave, Mole Creek, Tasmania (see plate 3.8).

Many major trunk cave passages that drain along the strike are within anticlines. The extension of joints in these locations promotes greater dissolution. In Kentucky, Mammoth Cave has its principal passages formed within the noses of plunging anticlines (Ford and Williams 1989, 41). Both Ogof Fynnon Ddu and Dan yr Ogof in Wales show anticlinal control (Lowe 1989; Charity and

Christopher 1977). Anticline Cave (Buchan, Australia) is a very small but clear example, while much of the hundred kilometres of Clearwater Cave, Mulu, Sarawak, is formed in an anticline.

Some examples of synclinal guidance over karst drainage are the Sistema Purificacion (Mexico; Hose 1981) and the Mamo-Atea Kananda (Papua New Guinea; Francis et al. 1980). In both cases streams sinking high on the flanks of the syncline have developed cave passages down dip which unite in the trough of the syncline.

The largest known underground chamber also owes its existence to a combination of folding and faulting. Sarawak Chamber in Lubang Nasib Bagus (Good Luck Cave) has a volume of approximately 12 million cubic metres, and floor dimensions of 600 m by 415 m (see figure 3.12). The cave is developed in upper Eocene to lower Miocene limestones which are partially recrystallized and folded. The large chamber has formed close to an anticlinal axis and to the underlying impermeable schists and slates (Gilli 1993) (see figure 3.13). There is also a fault on the western side of the chamber. Stoping of the chamber is aided by fretting and rotation of blocks in the arched roof, and the fallen blocks may be rapidly removed by stream action. The large passage of nearby Deer Cave has formed in a similar geologic context.

The deepest known cave, Réseau Jean-Bernard (−1602 m, Haute-Savoie, France), is developed on the flank of a syncline formed in

Plate 3.8
The roof of Marakoopa Cave, Mole Creek, Tasmania, has developed along a fault plane with slickensides.

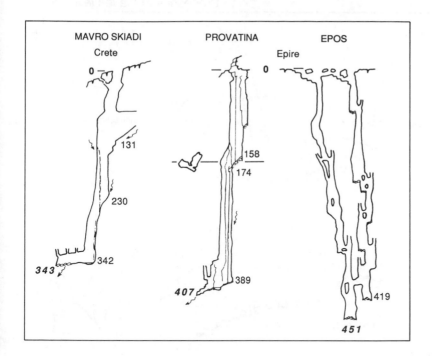

MAVRO SKIADI
Crete

PROVATINA
Epire

EPOS

Figure 3.11
*Long profiles of deep pits of
Crete and Epirus, Greece.
From Courbon et al. (1989).*

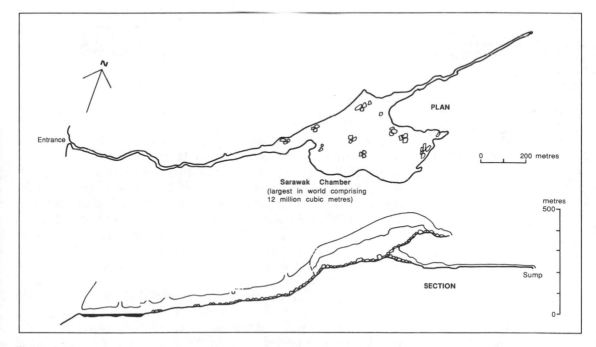

PLAN

Entrance

Sarawak Chamber
(largest in world comprising
12 million cubic metres)

0   200 metres

metres
500

SECTION

Sump

0

Figure 3.12
*Plan and section of Lubang Nasib Bagus (Good Luck Cave), Mulu, Sarawak, showing Sarawak
Chamber. From Courbon et al. (1989).*

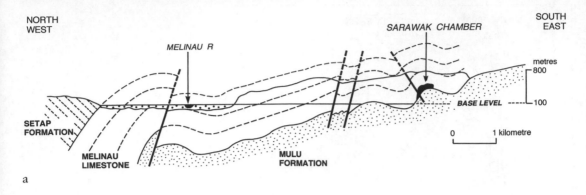

Figure 3.13
*Geological context of
Sarawak Chamber. The
chamber lies on a minor
anticlinal axis and has
been affected by faulting of
the Melinau Limestone.
From Gilli (1993).*

Lower Cretaceous Urgonian limestone (Lips et al. 1993). The cave's
upper entrances are found between 1800 and 2300 m in glaciated
karst, while its spring is found at 780 m in the Clevieux River. Karst
water has drained down the flank of the syncline, with an older
upper network of phreatic tubes graded to a former spring at
around 1800 m and a lower, younger meandering passage draining
to the present spring (see figure 3.14). The former spring is dated
from the regional denudation chronology as early Pliocene, at
which time valley incision was modest. Subsequent valley incision
has lowered the springs, with passage development enhanced by
glacial runoff. The change in passage morphology coincides with
an overthrust sheet of limestone on the Montagne du Criou;
some retardation of down-dip drainage may thus be impli-
cated. Fluvioglacial sediments in the cave have a minimum age of
*c.*190 000 years BP (Before Present) by uranium series dating; this
corresponds to the Mindel glaciation.

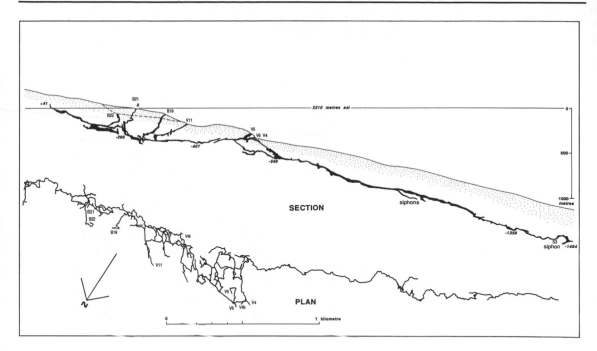

Cave breakdown and evaporite weathering

Limestone is a strong but somewhat brittle rock which fractures readily. The creation of breakdown in caves will be considered further in Chapter 5. Here the role of breakdown in shaping passage morphology will be reviewed. The role of hydrostatic support for cave cross-sections is evident when water-filled passages are drained; inevitably some wall and roof collapse occurs, causing a stoping process which continues through the beds until a stable arch is formed or a stronger interbed is reached. Valley-side unloading following hillslope erosion and fluvial incision may create a series of joints parallel to the valley axis. Close to cave entrances, this may produce a denser joint network which promotes collapse. Seismic shocks may also promote spalling of walls and opening of joints. Thus in any cave evolution sequence, ceiling and wall breakdown are an almost inevitable phase which may lead to partial or total infilling of the cavity.

The process of rock weathering by gypsum and halite crystallization (known as exsudation) can be a powerful agent for passage modification. Seepage water charged with soluble salts evaporates upon reaching the cave wall, and the expansion of crystals in bedding or small fissures promotes dramatic spalling.

There are good examples from the Nullarbor Plain of Australia, where solutional conduits, formed under phreatic conditions and then drained, have been so modified by exsudation as to remove

Figure 3.14
Section and plan of Réseau Jean-Bernard, Haute-Savoie, France. New passages giving a total depth of 1602 m are not shown. From Courbon et al. (1989).

nearly all the initial solution features and speleothems (Lowry and Jennings 1974; Gillieson and Spate 1992).

The Development of Common Caves

Formation of caves in plan

In Chapter 2 the factors governing the initiation of water circulation in a limestone massif were discussed. From this, the path of groundwater flow will depend on the balance between the direction in which resistance to flow is minimal and the direction of maximum potential energy (the shortest and steepest route from streamsink to spring). The long gradient of a karst hydrologic system will depend on the hydraulic head and the position of the springs in the system, the latter being controlled largely by geologic structure and base level. Once flow has been initiated, then a cave system will develop linking streamsinks and springs. The simplest situation is in the direction of steep dip. Initially a distributary pattern of solutional conduits develops, extending away from the water entry point down the hydraulic gradient (see figure 3.15). The precise pathways will depend on local lithology and structure, but in every case one tube will expand at the expense of the others and come to dominate the array. This will propagate down-gradient until the streamsink is linked to the spring. Once this occurs, resistance to flow is drastically reduced and a cave conduit

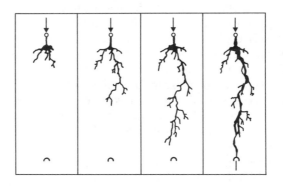

Figure 3.15
Development of cave conduits from a single input according to Ewers (1978). The conduit propagates from sink to spring by preferential development of one proto-conduit; the enlargement of this is shown in the lower detailed view.

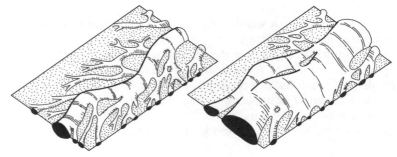

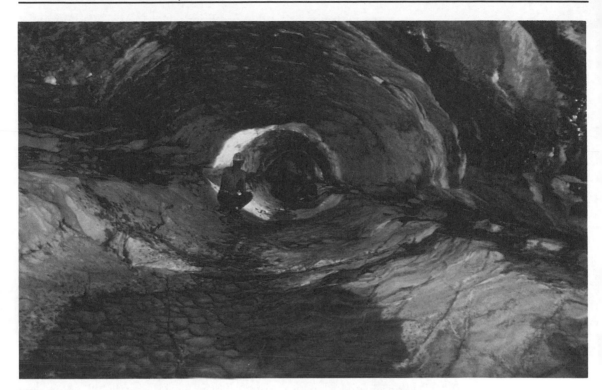

exists. Ford (1965) termed these dip tubes, but they may not follow the dip precisely, and Ford and Williams (1989, 253) employ the term *primary tubes* (see plate 3.9) for these features. Where the limestone massif is fractured, then the primary tubes will extend along fractures or bedding planes, zig-zagging to minimize flow resistance. Additional water inputs at joint or joint-bedding intersections will enhance the development of the cave conduit through the process of mixing corrosion. Thus the plan form of the developing cave network will vary between a simple, meandering tube (in the absence of dominant structural control) to highly angular or linear conduits where the rock is highly fractured. A more common situation is where there are multiple streamsinks draining to a spring or series of springs located at a base level determined by valley incision or geologic structure. Under this circumstance competitive development occurs, where one of the single input conduits links to and captures the flow from one or all of its neighbours. This capture occurs because flow resistance is not uniform in the rock mass, conduits form at different times, and pressure heads will not be equal. If one of the primary tube systems is close to the output boundary, then the local hydraulic gradient may be greater and capture may not occur. Thus there is complexity in plan form of karst hydrologic systems which depends on nuances of geologic structure and timing of events. Some examples of simple down-dip

Plate 3.9
A classic phreatic tube in Swine Hole, Peak Cavern, Derbyshire, UK. Photo by John Gunn.

systems with phreatic tubes are Friar's Hole Cave, West Virginia (see figure 1.2) and at Peak Cavern, Derbyshire, the Kingsdale Master Cave and Lost John's in the UK. The Subway in Castleguard Cave, Canada, is one of the best linear tube examples yet known. Finally, a classic high-dip example is provided by the Holloch, Switzerland.

Formation of caves in length and depth

So far the case of cave development along a single bedding plane has been investigated. Now attention is focused on the more normal case where water entry occurs along multiple bedding planes or fissures, introducing the possibility of complex development in dimensions of length and depth.

Ford and Williams (1989, 261) have proposed a four-state model (see figure 3.16) for the continuum of morphologies from deep phreatic to water table caves. In bathyphreatic or deep phreatic caves there is a single downward loop below the water table owing to high hydraulic resistance. There may be several elements to the loop, following the linking of conduits developed in separate bedding planes. At the inflow end there may be a drawdown vadose cave passage on account of the locally greater hydraulic gradient. As seen in Chapter 2, we now have good examples of water-filled bathyphreatic passages thanks to divers (see figure 2.4). At the Fontaine de Vaucluse, France, divers have descended 273 m in one such loop without reaching its point of inflection. When these passages are drained, their lower limbs are often obstructed by ceiling collapse or sediment fills, limiting exploration and mapping. Drilling for geological exploration in limestone has frequently encountered meteoric waters at great depths, but the precise nature of the conduits containing the water is unknown. In the Kutubu oilfield of Papua New Guinea such water has been encountered at a depth of 600 m. Dip tubes may be quite circular or more commonly elongated along joints or bedding. Flat roof sections owing to bedding control of dip tubes may also be seen in Dow Cave, Easter Grotto, Easegill Cave and Hasleden Cave (all in the UK), and in Roppel Cave, Kentucky.

According to Ford and Williams (1989, 263), where there is a higher fissure frequency, multiple loop phreatic caves occur. In such situations the piezometric surface is initially higher in the rock mass, but as the network develops and enlarges it is lowered until it roughly coincides with the top of the loops. In the Holloch, where such an array of passages has developed along irregular strike passages (Bögli 1970), the amplitude of the looping is about 100 m, with a maximum value of 180 m. In the reduced relief of the Kentucky sinkhole plain the amplitude is about 40 m.

With the diminished resistance to flow found in highly fissured rocks, a cave with a mixture of phreatic loops and water table-levelled elements may be found. Commonly such systems develop along the strike or along major joints, and they are a very common

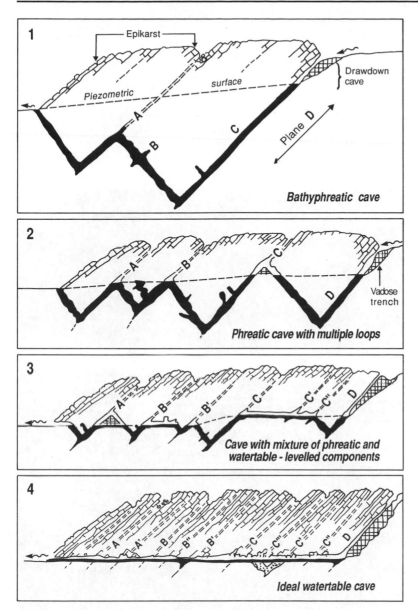

Figure 3.16
The four-state model differentiating the basic types of phreatic and water table caves. From Ford and Williams (1989).

type of cave. Multiple shallow loops, developed along individual bedding planes, are linked by short horizontal sections. In the Mendip Hills of England, the Swildon's–Wookey system has at least twenty such loops which today are sumps (see figure 2.4). Upstream exploration is at present halted at a depth of 80 m in such a loop. The passages of Clearwater Cave, Mulu, Sarawak, provide a good example of such a system which has been partially drained. There are numerous phreatic tubes at varying levels in the cave – for example, the Revival series, linking horizontal passages developed

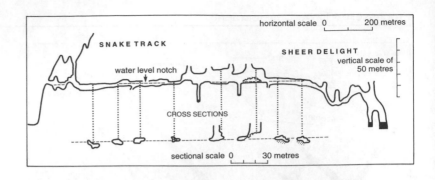

Figure 3.17
Wall notches in a section of Clearwater Cave, Sarawak, mark former stream levels in this multi-level phreatic and water table-levelled system. From Smart et al. 1985.

Within the figure:

horizontal scale 0 — 200 metres

SNAKE TRACK

water level notch

SHEER DELIGHT

vertical scale of 50 metres

CROSS SECTIONS

sectional scale 0 — 30 metres

Plate 3.10
The abandoned phreatic tube of Selminum Tem, Papua New Guinea, is 30 m in diameter and carried flood flows from the flanks of the glaciated Star Mountains at the height of the last Ice Age.

along the strike. In some sections of the cave, deep river incuts mark former water table positions, and may be temporarily invaded by floodwaters today (see figure 3.17). In Selminum Tem Cave, Papua New Guinea, abandoned phreatic tubes up to 30 m in diameter (see plate 3.10) are now perched several hundred metres above the piezometric surface feeding the Kaakil spring (Gillieson 1985).

Minor lowering of the piezometric surface will promote entrenchment of the top of the phreatic loops, giving short sections of vadose canyon (see plate 3.11 and figure 3.18). Bypass passages develop where blockage of a downward loop by detritus occurs (Renault 1968). This phenomenon is very common in tropical caves subject to flash flooding, where large quantities of fine-grained sediment are washed into loops and create a large hydraulic head across the loop. Fissures are opened to form short bypasses which may propagate downstream. Sometimes the downward loop may be re-excavated (as in Clearwater Cave) to create multiple flow paths dependent on flood stage.

Renault (1968) described a phenomenon termed *paragenesis* where the accumulation of sediment in a phreatic passage promotes solutional erosion at the top of the conduit (see figure 3.18) as the water is perched on top of the fill. The convergence of form between such features and normal vadose canyons makes determination of their origin difficult, and few examples cited so far are unequivocal. Recently Lauritzen and Lauritzen (1995) have developed a method based on a stereographic analysis of meander migration vectors developed from solutional scallop orientations. This innovation needs testing in a wider range of contexts but is very promising.

Finally, the ideal water table caves with long horizontal passages represent a fourth cave state developed where fissure frequency is high. Direct routes can be excavated between numerous input tubes developed down dip, and with enlargement these passages can absorb all of the flow. The piezometric surface lowers, yielding canal passages penetrable for hundreds of metres. These may flood to the roof following heavy rain, at which times some localized dynamic phreatic development may occur. This commonly results in flat roof sections cut across the bedding. Many of the long river caves in tectonically mobile regions are of this type. Good examples are provided by the Chuan Yan Cave of Nanxu, Guanxi, China (Waltham 1985); the Caves Branch system, Belize (Miller 1982); and Tham Nan Lang Cave, Thailand (Dunkley 1985), which is so large as to be explorable on elephant back for the first few hundred metres! Many of these caves have deep sediment fills, tens of metres thick, in both active and inactive passages. Cave streams may meander across the top of these sediment fills, eroding deep wall incuts which may subsequently be fully exposed by channel lowering.

It should be borne in mind that this simple four-state scheme of cave morphology represents a continuum, with individual caves

Plate 3.11
The entrance pitch of Ana Ahu Cave, Eua, Tonga, is typical of invasion vadose caves. Photo by John Gunn.

often exhibiting more than one of the states at a given time. It does not apply as well to those caves where water flow is localized, on account of rainwater inflow in a relatively small impounded karst. Such caves may show both phreatic and vadose passage forms at the same stage of development, and there may be a temporary phreatic state following intense rainfall. Such is the case for the extensive caves in dissected towers of the Chillagoe–Palmer River karst of north Queensland, and for many small caves in other tropical tower karsts of Malaysia and Thailand.

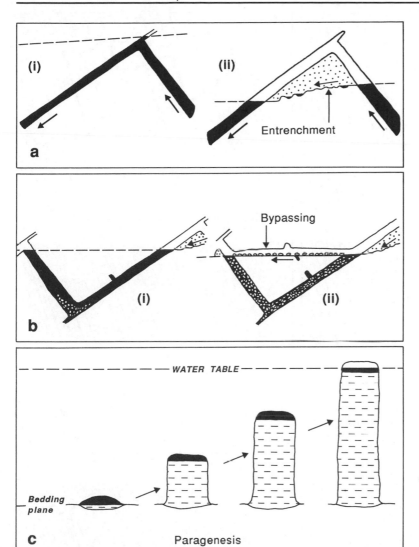

Figure 3.18
Common gradational features in phreatic caves: (a) isolated vadose entrenchment at the apex of a phreatic loop; (b) formation of a bypass tube across the top of a sediment-clogged phreatic loop; (c) upward development of a paragenetic passage to the water table. From Ford and Williams (1989).

According to Ford and Williams's (1989) model, fissure frequency exerts a dominant control over the long gradient of caves and the state of water flow. But fissure density evolves through time through the solutional enlargement of joints, providing a further possibility for changes in passage morphology. This is illustrated in figure 3.19, for a steeply dipping limestone. Fissure frequency diminishes markedly with depth in most limestone massifs, though exposed joints in cliffs, dolines and quarries are but a poor sample of the total. From an initial low-frequency state, increased density of linked fissures allows for the development of a bathyphreatic (type 1) cave system with deep loops. This may develop multiple loops and connections of lesser amplitude along enlarged joints and

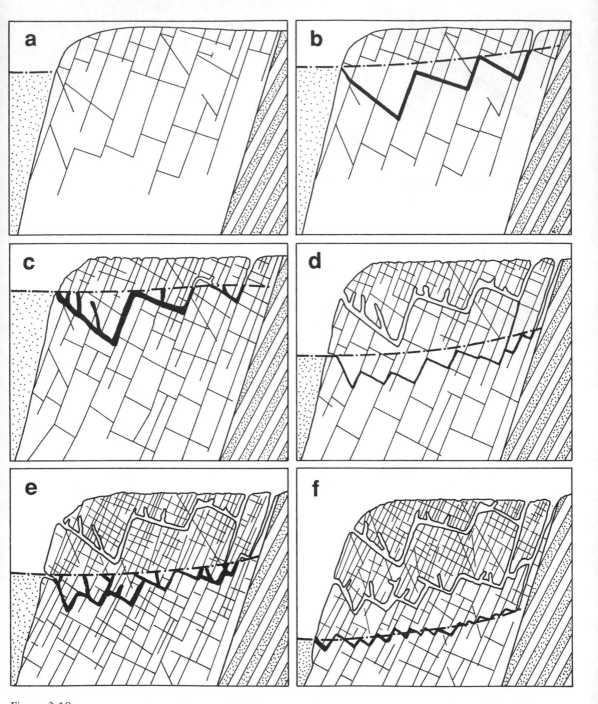

Figure 3.19
The geometry of successive caves in a multiphase system is affected by the increase in fissure frequency over time. First-generation caves are of the state 2, bathyphreatic type, while later caves tend to state 3 (mixed phreatic and water table) and 4 (water table-levelled) types. From Ford and Williams (1989).

fissures, giving a transition to types 2 and 3 as the spring lowers and upper passages are abandoned. Finally, in a highly fissured massif, a cave with a mixture of water table-levelled and phreatic passages may form. Such sequences are seen in the caves of Mulu, Sarawak, and in the caves of New Guinea (Smart et al. 1985; Farrant et al. 1995; Francis et al. 1980).

The Formation of Maze Caves

Caves formed by floodwaters

Maze caves often form where flat-lying well-bedded limestones are invaded by floodwaters (Palmer 1975). The juxtaposition of strike belts of limestone and river meanders creates a situation where water readily passes underground and erodes complex networks of intersecting passages, such as Sof Omar, Ethiopia, Bullock Creek, New Zealand (Crawford 1994), and the Atea Kananda, New Guinea (Francis et al. 1980). These act as meander cut-offs and may be found at several levels in an outcrop. Careful inspection of lithology may help to resolve whether multiple levels reflect changing base level control or exploitation of only the more soluble beds in the limestone. When carrying capacity of the main cave passage is reduced by sedimentation (both inorganic and organic), backflooding may occur creating temporary ponds. Large hydraulic heads may develop during floods, enhancing the enlargement of alternative flow paths. Plate 2.7 shows the backup of the Tekin River sink in New Guinea; during floods a lake 10 to 15 m deep forms at the entrance of this maze cave. A similar situation has been reported for the Baliem River sink of Irian Jaya (Checkley 1993).

Caves formed by hydrothermal waters

Maze caves may also form by hydrothermal action where either carbon dioxide-enriched waters or corrosive sulphuric acid derived from pyrites excavate well-jointed limestones.

The first case, where rising waters charged with carbon dioxide dissolve the limestone, is reasonably common and, according to Palmer (1991), accounts for about 10 per cent of caves known and explored. Water circulating deep in the rock mass becomes heated and gains carbon dioxide gas from deep sources. This heated water allows strong thermal convection; as the water rises it cools and the carbon dioxide gas dissolves. Three distinct types of cave formed by this mode are recognized:

- Rising hot water dissolves limestone upwards from a basal chamber along a branchwork of rising passages, which terminate in cupola-form dissolution pockets formed by slowly eddying waters. The best example is Satorkopuszta Cave in Hungary (Muller and Sarvary 1977).

- Water ascending fissures in limestone is halted by an aquiclude which compels the water to spread laterally, forming a two-dimensional maze or network, often joint guided. In some cases the hot water mixes with cooler groundwater and solution is enhanced by mixing corrosion. One example is Crossroads Cave, Virginia (Palmer 1991).

- By far the most common type is a three-dimensional maze formed in a single phase by rising warm water. Usually such caves are associated with modern hot springs discharging at stratigraphic boundaries with overlying or underlying non-karstic rocks, or along faults. The relict caves (Baradla system) of Buda Hill, Budapest, are of this type. The longest examples are Jewel Cave and Wind Cave of South Dakota, USA, where nearly 200 km of complex passages lie beneath the Black Hills (Ford 1985). They are formed in up to 140 m of well-bedded limestones capped by sandstones and shales. Modern hot springs discharge where this shale cap is breached by valley incision. These caves have little or no relation to the modern topography, and consist of large passages up to 20 m high encrusted in calcite spar speleothems with silica overgrowths. Isotopic study of these formations has shown them to be clearly of thermal origin (Bakalowicz et al. 1987). Regionally heated water from the granitic core of the Black Hills converged and ascended through the limestone prior to discharging at palaeosprings now removed by denudation.

The second case, that of limestones dissolved by waters containing hydrogen sulphide, has produced some of the best decorated caves in the world. Sulphuric acid can form by the oxidation of hydrogen sulphides derived from sedimentary basins, often associated with hydrocarbons. These sulphide-rich fluids can be released either slowly by the deep circulation of meteoric water, or rapidly by the migration of brines into marginal areas by tectonism, igneous activity or sediment compaction (Palmer 1991; Ford 1988). The caves of the Guadelupe Mountains, including the well-known Carlsbad Caverns, display a network appearance with great rooms formed in the reefal limestone linked to blind shafts and higher maze passages. The large chambers reflect zones of enhanced solution near the water table, and may have formed around existing vugs in the rock. Narrow fissures in their floors, clogged with speleothems, may be the original water entry points. The blind shafts represent the base of the mixing zone where sulphide-charged waters rose to meet meteoric waters. Flow of this acidified water into intersecting fissures has produced the maze passages.

If the concentration of dissolved sulphate is great enough, then extensive gypsum formations may result, as in Lechuguilla Cave, New Mexico (Davis 1980; Speleo Projects 1991). This is arguably

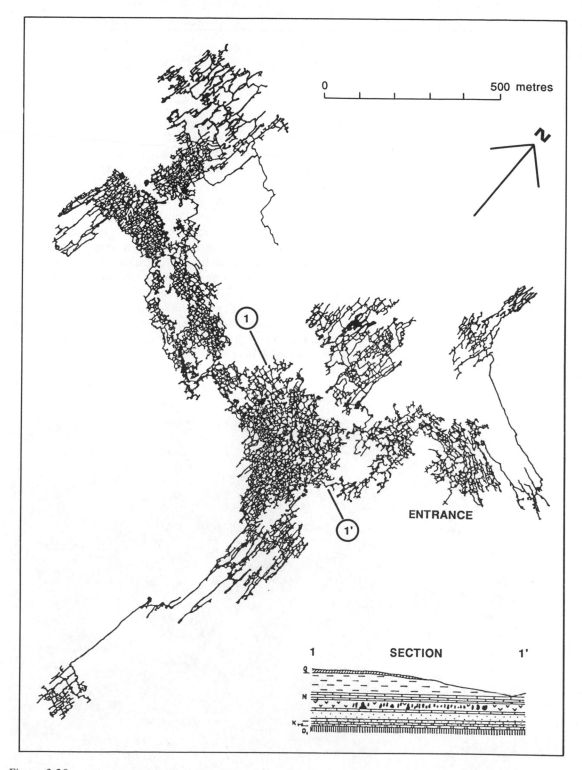

Figure 3.20
Plan of Optimisticheskaja Cave, Ukraine. This 165 km long maze cave is formed in gypsum strata separated by aquicludes of clay. From Courbon et al. (1989).

one of the most beautifully decorated caves in the world, containing a great diversity in forms and mineralogies of speleothems.

Caves formed in gypsum

Two of the world's longest caves are formed in the Upper Tertiary gypsum of western Ukraine. Each is well over 100 km long, and the two caves are separated by only 750 m horizontally. They are both maze caves (see figure 3.20) with a very high passage density. The caves are established in gypsum beds ranging from 10 to 40 m thick and have formed at two or three levels controlled by insoluble clay layers; in plan all the passages are controlled by joints which are perpendicular to each other (see plate 3.12). This jointing is a consequence of block faulting on a transition slope between the East European platform and the Pre-Carpathian foredeep. They are uniformly elliptical tubes, elongated along the bedding, and are typically 3 to 10 m wide and 3 to 8 m high.

Two theories have been advanced to account for their formation. Originally it was thought that intermittent surface streams sank into the gypsum and formed caves of shallow phreatic and water table origin controlled by the jointing (Dubljanksy 1979; Jakucs and Mezosi 1986). A new theory (Klimchouk 1992) proposes the upward development of the tubes from phreatic passages at the

Plate 3.12
A typical passage in the gypsum maze cave Ozernaja, Ukraine. Photo by John Gunn.

base of the gypsum, via discharge outlets for groundwater linking lower and upper passages. There are thousands of these outlets distributed uniformly along the master passages, and the evidence to date supports the latter theory.

Lava Tubes, Weathering Caves and Pseudokarst

The formation of lava tubes, weathering caves and sandstone caves is another aspect of the subject where the dividing line between true solution and disintegration of the rock may be hard to define. Caves in rocks other than limestone are widespread and sometimes are very large. There may be gross similarity in form although the processes leading to this are quite different.

The formation of lava tubes

Volcanic lava has a number of flow styles which depend on the viscosity of the molten rock. Basaltic lavas flow readily, and an internal network develops which conveys the liquid rock from the erupting vent to the advancing toe of the lava flow. This may be likened to an arterial system, with tributary branches, distributaries and anastomoses. This flow style is known as tube-fed pahochoe (Swanson 1973), and it is identified in solidified flows by the ropy surface and lobate structure leading to surface mounds or tumuli. Lava caves, the drained segments of lava tube networks, are part of this landform suite. The precise mode of formation of the lava caves was a matter of some controversy until observations were made during the 1969–74 eruptions of the Mauna Ulu vent of Kilauea volcano, Hawaii (Peterson and Swanson 1974). They confirmed that many of the large feeder tubes start as open lava canyons (similar to stream channels). Although the molten lava is at a temperature between 1000 and 1200°C, the surface of the flow loses heat rapidly to the air and forms a surface crust adhering to the banks. This crust may extend to bridge the flow completely, or fragments may detach and travel downflow until they jam across the channel. In another mode, liquid lava splashing and spattering near the vent may build inward-leaning levees which eventually meet to bridge the flow. Irrespective of the precise mechanism of bridging, these roofs are strengthened by further channel overflows until they stabilize. Tube-fed pahoehoe flows advance by the overlapping of successive small tongues, fed from the main axial mass of the molten lava. Smaller tubes or canyons may form as distributaries, from which small lava tongues develop at the edges of the flow or on its surface. Thus in plan there will usually be a main axial or feeder tube, many tens of kilometres long, which has a large diameter. These large tubes are sinuous and may be braided; there are numerous terrestrial examples as well as extra-terrestrial examples from Mars (Baker 1981). At the distal end of the flow, there is usually a network of braided distributary tubes and channels, similar to a delta. Here the rapid heat loss causes clogging of the distributaries, frequent overflows and development of new tubes and channels.

Complex and extensive caves are usually associated with the drainage of the axial feeder tube. These tubes are full of lava only during periods of high discharge, as lava delivery from the vents is pulsatory. Long continued flow may cause channel erosion, in a manner akin to water in stream channels, lowering the liquid level. The insulating properties of surrounding basaltic rock will permit some flow after vent activity has ceased, allowing lava flow into lower tubes and channels in a layered process. This sub-crustal drainage (Ollier 1977) may produce isolated cave segments, sealed above and below, which may be entered only after roof collapse has taken place (see plate 3.13). Liquid elements in a largely cooled and solidifed flow may exploit junctions or laminae between successive flows and enlarge them by remelting, producing further tube systems. Examples of this type are the multi-level lava caves of Mount Hamilton and Byaduk, western Victoria (Joyce 1980). Sagging of the roof may also produce surface depressions, linking cave entrances, which in plan give an idea of the form of the original axial feeder tube.

Currently the longest and deepest lava tube in the world is Kazumura Cave in Hawaii (Halliday 1995), which has a depth of 888 m and a length of 47 200 m. It extends over 28 km down flow and is a complex multi-level cave of the type described above. Its nearest competitor is Cueva del Vento in the Canary Islands, which is 478 m deep and 9250 m long.

Rarely lava flows may invade solution cavities in limestone. In Australia examples of *in situ* lava infilling have been described at Timor caves, in the upper Hunter valley (Connolly and Francis 1979) and at Bunyan, in southern New South Wales (Osborne 1979). These palaeokarsts are given a minimum age by the basalts, which date from the mid-Tertiary.

Weathering caves and pseudokarst

Pseudokarst processes involve the non-solutional erosion of rock to produce cavities, which may be isolated voids or connected passages or tubes, or surface karst features such as enclosed depressions and minor sculpturing of rock surfaces to resemble *karren*. Solutional features in non-carbonate rocks, such as sandstone, gypsum and halite, are excluded from this definition. However, solution can assist pseudokarst processes, so the distinction should be made according to the dominant process suite.

According to Grimes (1975), true pseudokarst involves the removal of material in the solid, liquid or gaseous state. These processes are summarized in table 3.5.

There are many examples of weathering caves in granite, laterite, and other granular rocks. Most of these are simple rock shelters,

Plate 3.13
Collapsed window into a lava tube, Surtshellir, Iceland.

though many have small tubes fed by seepage moisture and developed along bedding planes or joints. The dominant process in these is granular disintegration in which the interstitial cement is removed by solution or by corrasion. Thus caves formed in sandstone may be considered to be true karst or pseudokarst, depending on the dominant mode of passage enlargement. The large cave systems formed in the Proterozoic sandstones of Venezuela attain depths of 370 m and lengths of 2500 m (see figure 3.9). In these caves there are clearly solutional features alongside features due to granular disintegration. Under ever-wet tropical conditions even the low

Table 3.5   Classification of pseudokarst (modified from Grimes 1975, 7)

	State of removed material	
Solid	Liquid	Gas
Piping in alluvium or regolith	Melt water in glacier caves	Ablation hollows in snow packs and glaciers
Hydraulic action in sea caves	Thermokarst owing to heated water	
Mining and burrowing in mine tunnels and by fossorial mammals		
Fault fissures formed by tectonic movements		
Rock shelters and tafoni formed by granular disintegration, wind erosion and gravity fall		
Boulder caves formed by removal of regolith by water and gravity fall		

equilibrium solubility of silica, perhaps aided by acidic drainage waters, can over a long time produce substantial solutional conduits. On a more modest scale, the ancient sandstones of northern Australia have many stream caves and a type of 'beehive terrain' which superficially resembles cone karst (Young 1986; Jennings 1985). Finlayson (1982) has described small stream caves developed in granite in southern Queensland, while Greenhorn Cave (USA) is a much larger granite cave (Courbon et al. 1989).

Origin of Caves: an Overview

One of the major controversies surrounding the formation of caves rests on the nature of the initial cavity or cavities which guided the flow of water dissolving the rock. The divergent views can be broadly summarized into three classes:

1   The *kinetic* view, in which the size of tiny capillaries in the rock determines whether flow is laminar or turbulent; in the latter case, the helical flow characteristic of larger tubes permits accelerated solution with positive feedback. Over time this results in the preferential enlargement of a single tube which dominates the array and becomes the principal cave conduit (Ford and Ewers 1978).

2   The *inheritance or inception horizon* view, in which a pre-existing small cavity or chain of vugs, formed by tectonic, diagenetic (mineralization) or artesian processes, is in-

      vaded and enlarged by karstic groundwater to form a cave
conduit (Lowe 1992).

3   The *hypogene* view, in which hydrothermal deep waters
(charged with carbon dioxide, hydrogen sulphide or other
acids) result in the formation of heavily mineralized cavities which may be invaded by cool karst waters to give
larger, integrated cavities or networks (Davis 1980;
Bakalowicz et al. 1987).

Each view has its adherents and detractors, depending on the cave
experiences of the individual. It is possible that a single cave system
may have formed under the operation of some or all of these
processes during its long history. Clearly when thinking about cave
development we must critically re-examine the evidence, make
careful observations and keep an open mind about timescales.

A major change in thinking about the origin of caves has resulted
from the discoveries made in the last decade by a growing number
of cave divers who have described cavities below the water rest
level. In air-filled caves with sediments we may be looking only at
a very small proportion of the evidence; excavations and seismic
analysis demonstrate that the depth of sediment may be tens of
metres. In addition, the realization that palaeokarstic cave forms
and infills are widespread has radically revised thinking about the
age of individual cave systems. Thus the antiquity of caves is no
longer constrained into the timeframe of the Quaternary, and it is
clear that many cave systems have existed in some form throughout
both the Quaternary and the Tertiary. In some cases relict cave
passages may be as old as the Mesozoic (Osborne and Branagan
1988). Thus consideration of cave development must take account
of tectonic and climatic change over timescales where our knowledge of the conditions obtaining is very fragmentary. Reliable age
estimates from radiometric and microfossil analysis may help to
constrain the age of the speleothems and sediments, but not necessarily the cavities which they infill.

Given these conceptual revisions, it now seems likely that any
cave system may have formed in a variety of ways at various times.
Elucidation of the sequence of cave formation must rest on careful
examination of morphology above and below the present water rest
level. However, the factors affecting cave formation may be summarized as presented in table 3.6.

The life history of a single cave may be envisaged as passing
through a sequence of stages, not necessarily irreversibly, from an
initial state of no conduits to a long inception state during which
certain horizons or inhomogeneities in the rock mass act to channel
water into preferred pathways. Following this a developmental
phase, in which conduits form and enlarge, will persist and will
respond to external factors of base level change, rock mass evolution and tectonism. Finally in a multi-level system passages will be
abandoned and will collapse, leading to a final state of no con-

Table 3.6 Factors affecting speleogenesis

Speleogenesis	Stage	Factor
	Inception	Structure
		Lithology
		Chemistry
	Development	Palaeokarst
		Hydrology
		Climate
		Timescale

nected, enterable caves at a particular level. The sequence may be idealized to:

- inception
- gestation or nothephreatic
- turbulent threshold
- conduit and cave development
- decay and abandonment.

Gestation implies a phase of slow growth prior to the creation of a conduit. During timescales of speleogenesis which span millions, tens of millions or even hundreds of millions of years, processes similar to those of today have operated and their products can be seen – either wholly preserved or in fragments (Lowe 1992). From these, process suites can be reconstructed. Equally other processes have operated to produce relict features which have no contemporary analogues. There is a problem of convergence of forms from differing processes as well. The only way through this speleogenetic maze is careful observation, appropriate dating and healthy skepticism.

Geological Control and the World's Longest Cave

Mammoth Cave, Kentucky, is currently the longest cave in the world, and displays elegantly the role of geology in determining passage shape and orientation. The cave system is located at the southeastern edge of the broad Illinois structural basin, in limestones of Mississippian (early Carboniferous) age which dip gently to the northwest. The cavernous limestones are underlain by impure limestones of similar age and overlain by Mississippian and Pennsylvanian (late Carboniferous) sandstone, shale and thin limestone.

The limestone forms a broad plain of low relief, the Pennyroyal Plateau, punctuated by dolines and dissected by several sinking streams draining to the Green River to the north. The border of the limestone plain is an upland (the Chester cuesta) composed of limestone ridges capped by sandstones and other insoluble rocks. The Mammoth Cave system lies under these ridges, which are separated by broad valleys. Mammoth Cave occupies a 110 m thick sequence of limestones (see figure 3.21), with the lowest levels in the St Louis Limestone containing chert beds and some gypsum. The Ste Genevieve Limestone in the middle of the sequence contains most of the cave passages, and is made up of interbedded limestone and dolomite with some silty beds in the upper half. The Girkin Formation at the top of the sequence is 20 to 40 m thick and consists of limestone with thin interbedded shales and siltstones.

Most canyons and tubes in Mammoth Cave are highly concordant with the bedding of the limestones (Palmer 1981). Thus the main passage leading in from the Historic Entrance follows the same beds of the Ste Genevieve Limestone for several kilometres. Passages tend to parallel the strata because there are very few joints, and they rarely intersect more than one bed. In the southeastern area of Mammoth Cave, abrupt changes in passage level are associated with faults and major joints. This continuity in passages makes it possible to correlate passages and their sediments on either side of valleys. For example, Great Onyx Cave is now isolated from the rest of the Mammoth Cave system (see figure 3.22), but its stratigraphic position and passage forms suggest that it is genetically linked to Salts Cave and Dyer Avenue in the Flint Ridge system, also located in the Paoli Member of the Girkin Formation. The large canyon of Kentucky Avenue can be traced stratigraphically in the Joppa Member of the Ste Genevieve Formation and correlates with Broadway and Gothic Avenue in the Historic Tour section of the cave.

The three most common passage types in Mammoth Cave are canyons, low gradient tubular passages and vertical shafts. The

Figure 3.21
Major passages in Mammoth Cave, Kentucky, and their relationship to the limestone formations. Passage levels tend to rise along a gradient away from the Green River. From Palmer (1981).

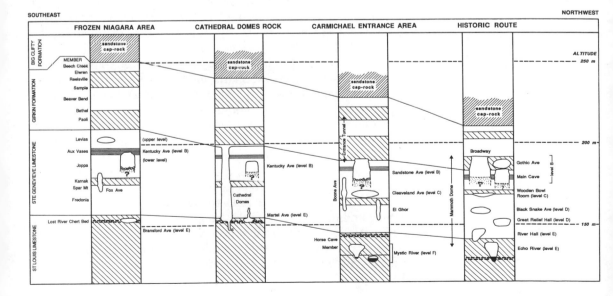

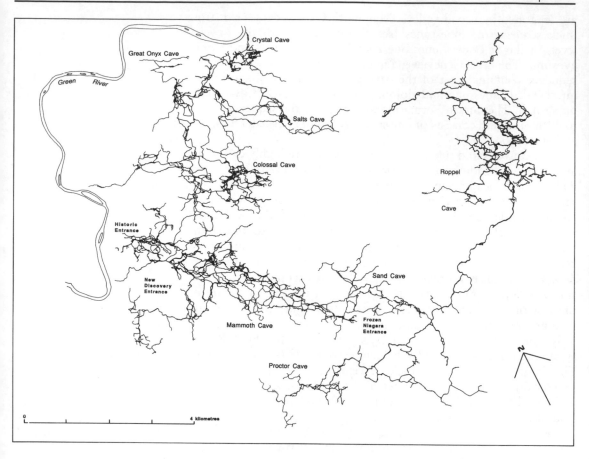

Figure 3.22
Plan of the Mammoth Cave system, Kentucky, including Colossal, Salts, Crystal and Roppel caves underlying Flint Ridge. From Courbon et al. (1989).

tubular passages are generally elliptical (see plate 3.14), reflecting the control of bedding (see figure 3.23). They trend along the strike and their sinuosity is controlled by local variations in dip and strike of the controlling bedding plane. The vadose canyons (see plate 3.15) are much modified by collapse but can be seen to drain down dip. The numerous vertical shafts in the cave are controlled by major joint intersections, with inflowing water perched on bedding planes or relatively resistant beds.

The dip tube of Cleaveland Avenue in Mammoth Cave shows geological control very well. It is about 1500 m long with almost no slope (4 m km^{-1}), and the limestone beds exposed in its walls show little variation along its length; it is clearly running almost exactly along the strike. Broad bends in the passage are caused by two gentle folds that push the tube to the west on the nose of an anticline and to the east into the trough of a syncline. It formed in the phreatic zone, and solution proceeded evenly at the junction of the Joppa and Karnak members of the Ste Genevieve Limestone to produce an elliptical tube with gently scalloped walls (see plate 3.16).

The concentration of large passages at certain levels is a combination of this stratigraphic control and the progressive downcutting of the Green River. Thus the passages cluster at three distinct elevations, 180, 168 and 152 m above sea level. Most passages at other elevations are shafts, narrow canyons and tubes. These passages have been backflooded by the river at the time when their levels were accordant, leaving the underground equivalents of floodplains and terraces. These deposits, and the speleothems that cap them, have provided a rich source of data for Quaternary scientists. The next section of this book explores the use of these cave deposits for environmental reconstruction.

Plate 3.14
The main passage of Indian Cave, Mammoth Cave National Park, Kentucky, an elliptical phreatic tube controlled by bedding. Photo by Gareth Davies.

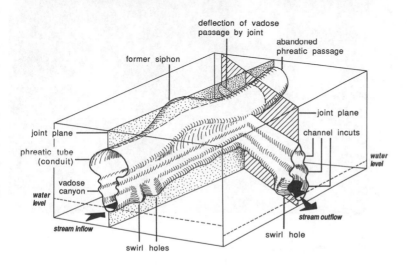

Figure 3.23
Principal morphological types of passages in Mammoth Cave, Kentucky, showing role of joint intersections. From Quinlan et al. (1983).

Plate 3.15
*Vadose canyon heavily
modified by collapse
processes at Dyer Avenue,
Crystal Cave, Kentucky.
Photo by F. D. Miotke.*

Plate 3.16
Large wall scallops in Colossal Cave, Flint Ridge, Kentucky. Photo by F. D. Miotke.

References

Back, W., Hanshaw, B. B. and van Driel, J. J. 1984: Role of groundwater in shaping the eastern coastline of the Yucatan peninsula, Mexico. In R. G. La Fleur (ed.) *Groundwater as a Geomorphic Agent*. Boston: Allen & Unwin, 281–93.

Bakalowicz, M. 1981: Les eaux d'infiltration dans l'aquifère karstique. *Proc. 8th Int. Congr. Speleo. Kentucky*, 2, 710–12.

Bakalowicz, M. J., Ford, D. C., Miller, T. E., Palmer, A. N. and Palmer, M. V. 1987: Thermal genesis of dissolution caves in the Black Hills, South Dakota. *Bull. Geol. Soc. Am.* 99, 729–38.

Baker, V. R. 1981: Pseudokarst on Mars. *Proc. 8th Int. Congr. Speleo., Kentucky*, 1, 63–5.

Bögli, A. 1964: Mischungskorrosion: ein Beitrag zum Verkarstungsproblem. *Erdkunde* 18 (2), 83–92.

Bögli, A. (ed.) 1970: *Le Holloch et son karst*. Neuchâtel: Baconnière.

Bögli, A. 1980: *Karst Hydrology and Physical Speleology*. Berlin: Springer, 116–213.

Charity, R. A. P. and Christopher, N. S. J. 1977: The Ogod Ffynnon Ddu cave system, South Wales, in relation to the structure of the Carboniferous Limestone. *Proc. 7th Int. Congr. Speleo., Sheffield*, 108–9.

Checkley, D. 1993: Cave of Thunder: the exploration of the Baliem River cave, Irian Jaya, Indonesia. *International Caver* 6, 11–17.

Connolly, M. and Francis, G. 1979: Cave and landscape evolution at Isaacs Creek, New South Wales. *Helictite* 17, 5–24.

Courbon, P., Chabert, C., Bosted, P. and Lindsley, K. 1989: *Atlas of the Great Caves of the World*. St Louis: Cave Books, 368 pp.

Crawford, S. 1994: Hydrology and Geomorphology of the Paparoa Karst, North Westland, New Zealand. Unpublished PhD thesis, University of Auckland, 240 pp.

Curl, R. L. 1986: Fractal dimensions and geometries of caves. *Math. Geol.* 18 (8), 765-83.

Davis, D. G. 1980: Cave development in the Guadelupe Mountains: a critical review of recent hypotheses. *Bull. Nat. Speleo. Soc. Am.* 42 (3), 42–8.

Dreybrodt, W. 1981: Kinetics of the dissolution of calcite and its applications to karstification. *Chem. Geol.* 31, 245–69.

Dubljansky, V. N. 1979: The gypsum caves of the Ukraine. *Cave Geol.* 1 (6), 163–83.

Dunkley, J. R. 1985: Karst and caves of the Nam Lang-Nam Khong region, Thailand. *Helictite* 23 (1), 3–22.

Ewers, R. O. 1978: A model for the development of broad scale networks of groundwater flow in steeply dipping carbonate aquifers. *Trans. Brit. Cave Res. Assoc.* 5 (2), 121–5.

Farrant, A. R., Smart, P. L., Whitaker, F. F. and Tarling, D. H. 1995: Long-term Quaternary uplift rates inferred from limestone caves in Sarawak, Malaysia. *Geology* 23 (4), 357–60.

Finlayson, B. L. 1982: Granite caves in Girraween National Park, southeast Queensland. *Helictite* 20, 53–9.

Folk, R. L. 1959: Practical petrographic classification of limestones. *Bull. Am. Soc. Petrol. Geol.* 43, 1–38.

Folk, R. L. 1962: Spectral subdivision of limestone types. *Am. Assoc. Petrol. Geol., Mem.* 1, 62–84.

Ford, D. C. 1965: The origin of limestone caverns: a model from the central Mendip Hills, England. *NSS Bull.* 27, 109–32.

Ford, D. C. 1985: Dynamics of the karst system: A review of some recent work in North America. *Annales de la Société Géologique de Belgique* 108, 283–91.

Ford, D. C. 1988: Characteristics of dissolutional cave systems in carbonate rocks. In N. P. James and P. W. Choquette (ed.) *Paleokarst*. New York: Springer.

Ford, D. C. and Drake, J. J. 1982: Spatial and temporal variations in karst solution rates: the structure of variability. In C. E. Thorn (ed.) *Space and Time in Geomorphology*. Boston: Allen & Unwin, 147–70.

Ford, D. C. and Ewers, R. O. 1978: The development of limestone cave systems in the dimensions of length and depth. *Can. J. Earth Sci.* 15, 1783–98.

Ford, D. C. and Williams, P. W. 1989: *Karst Geomorphology and Hydrology*. London: Unwin Hyman.

Francis, G., Gillieson, D. S., James, J. M. and Montgomery, N. R. 1980: Underground geomorphology of the Muller Plateau. In J. M. James and H. J. Dyson (ed.) *Caves and Karst of the Muller Range*. Sydney: Speleological Research Council, 110–17.

Friederich, H. and Smart, P. L. 1981: Dye tracer studies of the unsaturated zone: recharge of the Carboniferous Limestone aquifer of the Mendip Hills, England. *Proc. 8th Int. Congr. Speleo., Kentucky*, 283–6.

Gibbons, J. R. H. and Clunie, F. G. A. U. 1986: Sea level changes and Pacific prehistory. *J. Pacific History* 21, 58–62.

Gilli, E. 1993: Les grandes volumes souterrains du massif de Mulu (Borneo, Sarawak, Malaisie). *Karstologia* 22, 1–14.

Gillieson, D. 1985: Geomorphic development of limestone caves in the Highlands of Papua New Guinea. *Zeitschrift für Geomorphologie* 29, 51–70.

Gillieson, D. S. and Spate, A. P. 1992: The Nullarbor karst. In D. S. Gillieson (ed.) *Geology, Climate, Hydrology and Karst Formation: Field Symposium in Australia: Guidebook*, Special Publication No. 4, Department of Geography and Oceanography, University College, Australian Defence Force Academy, Canberra, 65–99.

Gillieson, D., Gorecki, P., Head, J. and Hope, G. 1986: Soil erosion and agricultural history in the Central Highlands of New Guinea. In V. Gardiner (ed.) *International Geomorphology*. London: Wiley, 507–22.

Grimes, K. G. 1975: Pseudokarst: definition and types. *Proc. 10th. Bienn. Conf. Aust. Speleo. Fedn, Brisbane*, 6–10.

Gunn, J. 1981: Limestone solution rates and processes in the Waitomo district, New Zealand. *Earth Surface Processes and Landforms* 6, 427–45.

Gunn, J. 1986: Solute processes and karst landforms. In S. T. Trudgill (ed.) *Solute Processes*. Chichester: Wiley, 363–437.

Halliday, W. R. 1995: 'Puna emergency road proposal – Kazumura Cave'. (http://www.halcyon.com/samara/nssccms/puna1.html) October 1995.

Hanshaw, B. B. and Back, W. 1979: Major geochemical processes in the evolution of carbonate-aquifer systems. *J. Hydrol.* 43, 278–312.

Hose, L. D. 1981: Fold development in the Anticlinorio Huizachal-Peregrina and its influence on the Sistema Purificacion, Mexico. *Proc. 8th Int. Congr. Speleo., Kentucky*, 133–5.

Jakucs, L. 1977: *Morphogenetics of Karst Regions: Variants of Karst Evolution*. Budapest: Akademiai Kiado.

Jakucs, L. and Mezosi, G. 1986: Genetic problems of the huge gypsum caves of the Ukraine. *Acta Geografica* 16, 15–38.

Jennings, J. N. 1967: Further remarks on the Big Hole, near Braidwood, New South Wales. *Helictite* 6, 3–9.

Jennings, J. N. 1968: Geomorphology of Barber Cave, Cooleman Plain, New South Wales. *Helictite* 6, 23–9.

Jennings, J. N. 1983: Sandstone pseudokarst or karst? In R. L. Young and G. Nanson (ed.) *Aspects of Australian Sandstone Landscapes*. Wollongong: Australian and New Zealand Geomorphology Group, 21–30.

Jennings, J. N. 1985: *Karst Geomorphology*. Oxford: Blackwell.

Joyce, E. B. 1980: Origin of lava caves. *Proc. 13th Bienn. Conf. Aust. Speleo. Fedn,* Melbourne, 40–8.

Klimchouk, A. B. 1992: Large gypsum caves in the Western Ukraine and their genesis. *Cave Science* 19, 3–11.

Lauritzen, S. E. and Lauritzen, A. 1995: Differential diagnosis of paragenetic and vadose canyons. *Cave and Karst Science* 21, 55–60.

Laverty, M. 1987: Fractals in karst. *Earth Surface Processes and Landforms* 12, 475–80.

Leighton, M. W. and Pendexter, C. 1962: Carbonate rock types. *Am. Ass. Petrol. Geol. Mem.* 1, 33–61.

Lips, B., Gresse, A., Delamette, M. and Maire, R. 1993: Le Gouffre Jean Bernard (−1602 m, Haute Savoie, Fr.): écoulements souterrains et formation du reseau. *Karstologia* 21, 1–14.

Lowe, D. J. 1989: Limestones and caves of the Forest of Dean. In T. D. Ford (ed.) *Limestones and Caves of Wales*. Cambridge: Cambridge University Press, 106–16.

Lowe, D. J. 1992: A historical review of concepts of speleogenesis. *Cave Science* 19, 63–90.

Lowry, D. C. and Jennings, J. N. 1974: The Nullarbor karst, Australia. *Zeitschrift für Geomorphologie* 18, 35–81.

Mangin, A. 1975: Contribution à l'étude hydrodynamique des aquifères karstiques. DES thesis, University of Dijon *(Ann. Spéléo.* 29 (3), 1974, 283–332; 29 (4), 1974, 495–601; 30 (1), 1975, 21–124).

Miller, T. E. 1982: Hydrochemistry, Hydrology and Morphology of the Caves Branch Karst, Belize. PhD thesis, McMaster University.

Muller, P. and Sarvary, I. 1977: Some aspects of developments in Hungarian speleology theories during the last ten years. *Karszt-es Barlang,* 53–9.

Nunn, P. D. 1994: *Oceanic Islands*. Oxford: Blackwell, 416 pp.

Ollier, C. D. 1975: Coral island geomorphology – the Trobriand Islands. *Zeitschrift für Geomorphologie* 19, 164–90.

Ollier, C. D. 1977: Lava caves, lava channels and layered lava. *Atti de Seminario sulle Grotte Laviche*. Catania: Gruppo Grotte Catania, 149–58.

Osborne, R. A. L. 1979: Preliminary report on a cave in tertiary basalt at Coolah, New South Wales. *Helictite* 17, 25–9.

Osborne, R. A. L. and Branagan, D. F. 1988: Karst landscapes of New South Wales. *Earth Science Reviews* 25, 467–80.

Palmer, A. N. 1975: The origin of maze caves. *Bull. Nat. Speleo. Soc. Am.* 37 (3), 56–76.

Palmer, A. N. 1981: *A Geological Guide to Mammoth Cave National Park*. Teaneck, NJ: Zephyrus Press, 196 pp.

Palmer, A. N. 1991: Origin and morphology of limestone caves. *Bull. Geol. Soc. Am.* 103, 1–21.

Palmer, R. J. 1989: *Deep into Blue Holes*. London: Unwin Hyman, 164 pp.

Peterson, D. N. and Swanson, D. A. 1974: Observed formation of lava tubes during 1970–1971 at Kilauea Volcano, Hawaii. *Studies in Speleology* 2, 209–22.

Picknett, R. G., Bray, L. G. and Stenner, R. D. 1976: The chemistry of cave waters. In T. D. Ford and C. H. D. Cullingford (ed.) *The Science of Speleology*. London: Academic Press, 593 pp.

Playford, P. D. and Lowry, D. C. 1966: Devonian reef complexes of the Canning Basin, Western Australia. *Bull. Geol. Surv. W. Aust.* 118.

Plummer, L. N. 1975: Mixing of seawater with calcium carbonate ground water: quantitative studies in the geological sciences. *Geol. Soc. Am. Mem.* 142, 219–36.

Renault, H. P. 1968: Contribution à l'étude des actions mechaniques et sedimentologiques dans la spéléogenèse. *Ann. Spéléo.* 22, 5–21, 209–67; 23, 259–307, 529–96; 24, 313–37.

Smart, P. L., Bull, P. A., Rose, J., Laverty, M. and Noel, M. 1985: Surface and underground fluvial activity in the Gunung Mulu National Park, Sarawak: a palaeoclimatic interpretation. In I. Douglas and T. Spencer (ed.) *Tropical Geomorphology and Environmental Change*. London: Allen & Unwin, 123–48.

Smith, D. I. and Atkinson, T. C. 1976: Process, landforms and climate in limestone regions. In E. Derbyshire (ed.) *Geomorphology and Climate*. London: Wiley, 369–409.

Speleo Projects 1991: *Lechuguilla: Jewel of the Underground*. Basel: Speleo Projects, 144 pp.

Swanson, D. A. 1973: Pahoehoe flows from the 1969–1971 Mauna Ulu eruption, Kilauea Volcano, Hawaii. *Bull. Geol. Soc. Am.* 84, 615–26.

Trudgill, S. T. 1976: The erosion of limestone under soil and the long-term stability of soil-vegetation systems on limestone. *Earth Surface Processes and Landforms* 1, 31–41.

Trudgill, S. T. 1985: *Limestone Geomorphology*. London: Longmans.

Truluck, T. 1994: The sandstone shafts of the Chimanimani Mountains. *Caves and Caving* 65, 15–18.

Waltham, A. C. (ed.) 1985: *China Caves '85*. London: Royal Geographical Society, 60 pp.

Wigley, T. M. L., Drake, J. J., Quinlan, J. F. and Ford, D. C. 1973: Geomorphology and geochemistry of a gypsum karst near Canal Flats, British Columbia. *Can. J. Earth Sci.* 10, 113–29.

Williams, P. W. 1983: The role of the subcutaneous zone in karst hydrology. *J. Hydrol.* 61, 45–67.

Williams, P. W. and Dowling, R. K. 1979: Solution of marble in the karst of the Pikikiruna Range, northwest Nelson, New Zealand. *Earth Surface Processes and Landforms* 4, 15–36.

Worthington, S. R. H. 1991: Karst Hydrogeology of the Canadian Rocky Mountains. Unpublished PhD thesis, McMaster University, 370 pp.

Young, R. W. 1986: Tower karst in sandstone: Bungle Bungle massif, northwestern Australia. *Zeitschrift für Geomorphologie* 30, 189–202.

Young, R. W. and Young, A. 1992: *Sandstone Landforms*. Berlin: Springer, 163 pp.

FOUR

Cave Formations

Introduction

Cave formations or speleothems are one of several kinds of cave interior deposits, the others being *in situ* breakdown materials and clastic sediments transported mechanically into and deposited within the cave (White 1976). Cave sediments, both coarse and fine, will be discussed further in Chapter 5. *Speleothems* are secondary chemical precipitates derived from cool or hot water circulating in the karst. These cave formations may be composed of only one or a combination of over two hundred known minerals which range in form from thin crusts a few millimetres thick to vertical columns 20 m high (see plate 4.1).

There is an enormous literature on the subject of cave formations. They have been the subject of review chapters (or parts thereof) in some key karst texts (e.g., White 1976, 1988; Bögli 1980; Bull 1983; Jennings 1985; Ford and Williams 1989) and have been described in great detail by Hill (1976) and Hill and Forti (1986), the latter of whom cite over 2000 references. Of the review works, perhaps the most comprehensive is that by White (1976). The part-chapter in Ford and Williams (1989) devoted to cave formations is broader in scope but less detailed. New cave minerals and cave formations are discovered every year, so any review will be quickly outdated. This is especially true in the humid tropics, where organic compounds derived from guano may combine with inorganic chemicals to produce a complex range of carbonate and phosphate minerals.

Carbonates

Carbonates are the most common mineral deposits found in caves (Hill and Forti 1986; Ford and Williams 1989); the ten secondary carbonate minerals listed by Hill and Forti (1986) are those associated with 'normal' karst waters (i.e., not derived from geothermal

Plate 4.1
Massive stalagmites in the
Hall of the Thirteen,
Gouffre Berger, France.
Photo by Gareth Davies.

water or from ore bodies). Of these, calcite is by far the most
important and, together with aragonite, constitutes perhaps 95 per
cent of all cave minerals. The incidence of other carbonates is
occasional or rare and usually confined to particular chemical
environments or environmental conditions (see table 4.1).

The main processes by which carbonate formations are deposited
are by CO_2 loss or degassing, and by evaporation. In the case of

Table 4.1 Carbonate minerals commonly found in caves (from Hill and Forti 1986)

Name	Formula	Crystal system	Distinctive properties
Aragonite	$CaCO_3$	Orthorhombic	Colourless or white Transparent to translucent Acicular habit: usually in the form of needles in radiating groups Common cave mineral Specific gravity determination: Aragonite – 2.95 Bromoform – 2.85 Calcite – 2.75 Calcite will float in bromoform while aragonite will sink
Calcite	$CaCO_3$	Trigonal	Usually colourless or white, but can be stained various shades of red, tan or grey Transparent to translucent Hardness: 3 Usually massive, but sometimes in the form of a scalenohedron or rhombohedron Extremely high birefringence The most common cave mineral Effervesces vigorously in cold dilute hydrochloric acid
Dolomite	$CaMg(CO_3)_2$	Trigonal	White, tan or pink Transparent to translucent Effervesces slightly in cold dilute acid Rare in caves Must be distinguished from calcite by optical or X-ray diffraction techniques
Huntite	$CaMg_3(CO_3)_4$	Trigonal	White Dull, powdery, very fine-grained Rare in caves X-ray techniques required for positive identification
Hydromagnesite	$Mg_5(CO_3)_4$ $(OH)_2 \cdot 4H_2O$	Monoclinic (pseudo-orthorhombic)	White Vitreous to silky when wet, earthy and powdery when dry. Feels like cream cheese when rubbed between the fingers

Table 4.1 *Continued*

Name	Formula	Crystal system	Distinctive properties
			Extremely fine-grained Rare in most caves A common constituent of moonmilk Cannot be positively identified without an X-ray examination
Magnesite	$MgCO_3$	Trigonal	White Tasteless Powdery and fine-grained Rare in caves Should be identified by X-ray techniques
Monohydrocalcite	$CaCO_3 \cdot H_2O$	Hexagonal	Colourless to white Very rare; forms as a cold-water aerosol product in alpine caves Identification must be by X-ray diffraction
Nesquehonite	$MgCO_3 \cdot 3H_2O$	Monoclinic	White Massive, fine-grained Easily soluble in dilute acid Rare in caves Should be identified by X-ray techniques
Siderite	$FeCO_3$	Trigonal	Yellowish-brown to amber Isomorphous with calcite As cores in spar
Vaterite	$CaCO_3$	Hexagonal	White High-temperature polymorph of calcite Found in carbidimite formations

CO_2 loss, percolation waters rich in dissolved carbon dioxide of soil origin enter the cave passage through fissures and pores (see figure 4.1). Once exposed to the cave atmosphere, which is relatively depleted in CO_2, the percolation waters will equilibrate with their new surrounds and CO_2 will be lost from solution. This raises saturation levels of the solution with respect to calcite (and, where applicable, other minerals in solution), creating the necessary preconditions for mineral precipitation. If suitable nuclei are available and kinetic barriers are overcome, calcite will inevitably precipitate (see plate 4.2).

Evaporation concentrates CO_2 and dissolved ions in the source waters; this also raises saturation levels and may trigger calcite precipitation. Degassing of CO_2 without the assistance of evaporation is considered to be the primary cause of carbonate deposition (White 1976). Calcite formed under evaporative conditions (usu-

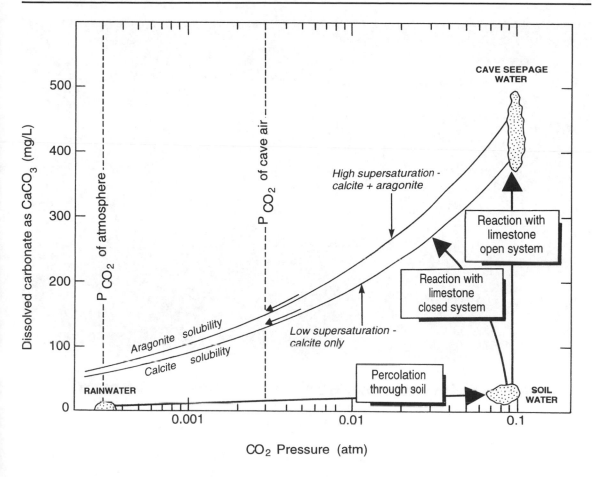

CO₂ Pressure (atm)

Figure 4.1
*Effects of pCO₂ variation
on the equilibrium
solubility of calcite and
aragonite in caves fed by
percolation water. Closed
systems tend to produce
calcite deposition alone,
while open systems
produce supersaturated
solutions depositing both
calcite and aragonite. From
White (1988).*

ally near cave entrances, in draughty caves and in caves in dry regions) is often microcrystalline and soft; by contrast, calcite formed by CO_2 degassing is usually hard and massively crystalline (Ford and Williams 1989). The most readily understood accounts of carbonate precipitation processes are provided by White (1976), Picknett et al. (1976) and Dreybrodt (1980).

The mechanisms by which calcite and other cave carbonates are formed has generated much interest over several centuries. Despite this, and notwithstanding the fact that the general processes of carbonate deposition are well understood, some variations on the above pedogenically derived CO_2 degassing and evaporation processes are worth noting. Such variations are important, for they may affect the mineralogy/polymorph or physical form of the deposit. For example, Dreybrodt (1982) considered the theoretical case of $CaCO_3$ deposition occurring in the absence of soil CO_2 so as to explain speleothem formation during glacial periods. Here, cold waters in equilibrium with atmospheric CO_2 percolate through bare or ice-covered karst and dissolve carbonate bedrock. The

Plate 4.2
Straw stalactites, Cloud Chamber, Dan-yr-Ogof, South Wales. Photo by Gareth Davies.

waters are warmed under closed-system conditions by their passage through the bedrock as well as upon entering the cave environment. Speleothems are thus deposited by CO_2 outgassing caused by temperature changes. This lack of biogenic input may explain the heavy $\delta^{13}C$ isotope signatures of some alpine cave deposits (Dreybrodt 1982; Harmon et al. 1983).

James (1977) showed that fluctuations in cave air CO_2 levels (rather than changes in the CO_2 of percolation waters *per se*) exert a major control on calcite deposition. Organic material washed into caves at Bungonia, New South Wales, during storm events and its subsequent decomposition by microorganisms contributed significantly to cave air CO_2 levels (over 5 per cent in places at various times) and served to control carbonate precipitation/dissolution cycles in deeper cave passages.

Atkinson (1983) noted the following complex dissolution/ mineral precipitation sequence from Castleguard Cave, Alberta, Canada:

1   dissolution of calcite and dolomite bedrock by carbonic acid (derived from atmospheric CO_2)
2   oxidation of pyrite to sulphuric acid and Fe hydroxides

3   further dissolution of calcite and dolomite by the sulphuric acid evolved from 2

4   precipitation of calcite following supersaturation owing to common-ion effects of 3.

An interesting version of the common-ion effect is invoked by Davis et al. (1990, 1992) in the development of subaqueous speleothems in Lechuguilla Cave, New Mexico. Odd worm-like tubes or tendrils are found growing beneath the margins of gour pools; capillarity cannot be used to explain these eccentric formations akin to helictites. Instead, the common-ion effect has been invoked. Small trickles of water enter the pool after passing over or between gypsum blocks. The dissolution of the gypsum boosts the calcium content of the water, and forces the less-soluble calcite to precipitate. The tubes or tendrils mark the entry points of the gypsum-charged water into the ponds.

Controls over carbonate mineralogy

Carbonate mineralogy is controlled to a large extent by ion ratios and pCO_2, according to Ford and Williams (1989). Given that calcite and aragonite are the most common cave minerals, it is not surprising that much effort has been spent in trying to understand the conditions under which each is formed. The best reviews of these precipitation processes are those of Curl (1962) and Hill and Forti (1986).

In thermodynamic terms, aragonite is 11 to 16 per cent more soluble than calcite (Harmon et al. 1983; Hill and Forti 1986). Thus, a solution which is just saturated with calcite is aggressive towards aragonite; under the temperature and pressure conditions likely to be encountered in caves, a given solution will preferentially precipitate calcite (Picknett et al. 1976). To this can be added the presence of a pre-existing calcite surface upon which further calcite growth is best suited compared with 'new' aragonite growth (Picknett et al. 1976).

Although its presence in speleothems was earlier thought to indicate warm conditions (e.g., Moore 1956), aragonite is found in a wide range of cave environments, including those high latitude alpine regions (Harmon et al. 1983). Thus, the idea that aragonite deposition is controlled by temperature is probably no longer tenable (Hill and Forti 1986); the principal reasons given for aragonite precipitation are the poisoning effect on calcite growth by foreign matter such as clays, and rapid CO_2 degassing at high supersaturation levels.

Foreign matter, such as magnesium ions, is thought to poison calcite crystal growth (White 1976; Bögli 1980; Bull 1983). Inhibition of calcite precipitation moves supersaturation levels into the range of aragonite precipitation. Numerous examples exist which document aragonite precipitating in preference to calcite in the presence of high concentrations of Mg^{2+} ions (e.g., Roques 1964; Cabrol and Coudray 1978; Gonzalez and Lohmann 1988). Cabrol and Coudray (1978), for example, found aragonite speleothems in southern France confined to those portions of caves directly beneath dolomite bedrock. Zeller and Wray (1956) found that strontium and lead initiated aragonite, while Bischoff and Fyfe (1968) noted that sulphate can inhibit calcite growth and favour aragonite precipitation. Zinc leached from galvanized piping may also inhibit calcite deposition around the paths of tourist caves.

Organic and/or clay particles have also been shown to be calcite inhibitors and thus indirect inducers of aragonite formation. In a petrographic study, Craig et al. (1984) observed that aragonite bands were initiated from clastic layers and solution boundaries in speleothems from Missouri. In laboratory experiments, they demonstrated that the addition of successively larger quantities of clay minerals into supersaturated solutions inhibited calcite growth and favoured the precipitation of metastable polymorphs (in this case vaterite). White (1976) mentioned the importance of adsorption of organic substances (especially amino acids) onto the steps and kinks of calcite crystals in blocking further calcite growth and promoting aragonite crystallization.

High levels of supersaturation and/or rapid degassing also induce aragonite precipitation. This is done most effectively via evaporation (Hill and Forti 1986), and in some ways evaporative and foreign ion effects are not mutually exclusive. In Carlsbad Caverns, New Mexico, Gonzalez and Lohmann (1988) found that evaporation played a significant role in aragonite precipitation. Aragonite was precipitated only after pore water Mg/Ca exceeded ~1.5; it was the dominant phase when the ratio extended beyond 2.5. High Mg/Ca contents were attributed to several factors. Initially, degassing caused precipitation of low magnesium calcite. This reduction in Ca^{2+} ions coupled with evaporation led to high Mg concentrations and the precipitation of aragonite followed by the precipitation of high magnesium calcite. Harmon et al. (1983) suggested that evaporation of Mg-enriched seepage waters triggered the precipitation of aragonite and other more soluble minerals in Castleguard Cave.

The incidence of the remaining carbonate minerals is occasional or rare. Hydromagnesite and other hydrated carbonates are microcrystalline aggregates formed under conditions of high supersaturation triggered by either evaporation and/or excessive degassing, and in most cases appear to be precipitated as residues (particularly as moonmilk) once thresholds of less soluble minerals have been breached (White 1976, 1982; Hill and Forti 1986). Dolomite, which is relatively common as a secondary deposit in subaerial environments (e.g., dolocretes), is quite rare in caves. Thrailkill (1968b) observed dolomite only as an alteration product of other metastable minerals in Carlsbad Caverns. However, Gonzalez and Lohmann (1988) found dolomite coating pool shelves and as layers within pool ledge deposits in the same environment. They ruled out an evaporative origin because other hydrated carbonates were absent; instead they suggested that dolomite may precipitate under conditions where pool water is in equilibrium with cave pCO_2 which becomes undersaturated with respect to calcite and aragonite but retains supersaturation with respect to dolomite.

Cave deposits formed by carbonate minerals

Reviews of carbonate cave formations are found in White (1976), Bögli (1980), Hill and Forti (1986) and Ford and Williams (1989). Initiation, development and the growth dynamics of basic soda straws, stalactites and stalagmites are dealt with by Moore (1962), Curl (1972, 1973), Cabrol and Coudray (1978), Franke (1975) and Gams (1981); these are summarized by Hill and Forti (1986), including speleothem growth rates (see table 4.2).

Ford and Williams (1989) list fourteen different conditions which may produce growth, erosion or stabilization of speleothem development. Petrological aspects are covered by Folk and Asserto (1976) and by Kendall and Broughton (1978). Varnedoe (1965)

Table 4.2 Measured growth rates of speleothems (modified from Hill and Forti 1986, 178–9); note that these are maximum rates for fast-growing formations

Mineral/ speleothem	Rate (mm/year)	Location
Calcite		
Stalagmite	0.05	Domica Cave, Slovakia
	0.17	Grotta Gigante, Italy
	0.25	Wyandotte Cave, Indiana, USA
	6.07	Clapham Cave, Yorkshire, UK
	7.66	Ingleborough Cave, Yorkshire, UK
Stalactite	0.06	Moaning Cave, California, USA
	0.15–0.20	Hand-Dug Cave, Kansas, USA
	0.17	Ingleborough Cave, Yorkshire, UK
	0.32	Postojna Cave, Slovenia
	1.00	Wyandotte Cave, Indiana, USA
	2.43	New Cave, Ireland
	3.20	Grotta Doria, Italy
Straw	0.21–0.45	Douchlata Cave, Bulgaria
	1.00	Rhine valley, Germany
	1.24	Han-sur-Lesse Cave, Belgium
	3.00–4.00	Slouperhöhle, Czech Republic
Crust	0.01	Postojna Cave, Slovenia
	0.17	Ochozerhöhle, Czech Republic
	0.25	Vypusterhöhle, Czech Republic
	0.36	Moaning Cave, California, USA
Gypsum		
Selenite crust	0.02	Mammoth Cave, Kentucky, USA
Selenite needles	0.07	Cove Knob Cave, West Virginia, USA

produced a hypothesis for rimstone dam formation which has
stood the test of time.

White (1976) categorized carbonate formations into three
groups:

1 Dripstone and flowstone forms:
   a) stalactites (see plate 4.1)
   b) stalagmites (see plate 4.2)
   c) draperies (see plate 4.3)
   d) flowstone sheets
2 Erratic forms:
   a) shields
   b) helictites (see plate 4.4)
   c) botryoidal forms (see plate 4.5)
   d) anthodites (see plate 4.6)
   e) moonmilk
3 Sub-aqueous forms:
   a) rimstone pools
   b) concretions
   c) pool deposits
   d) crystal linings.

Plate 4.3
*Shawl formation 250 cm
high in Junction Cave,
Wombeyan, New South
Wales.*

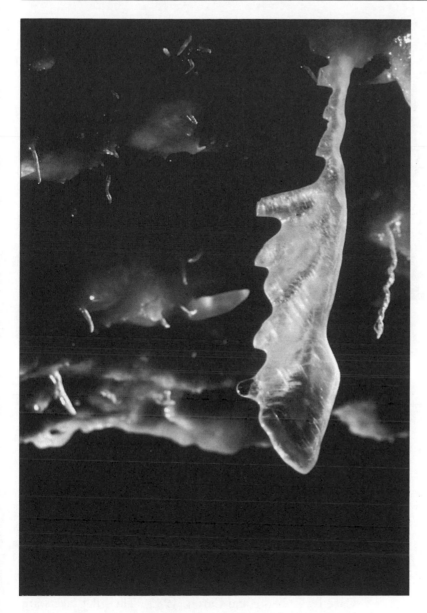

Plate 4.4
*Helictite formation in
Barellan Cave, Jenolan,
New South Wales. Photo
by Andy Spate.*

Downward-growing stalactites are formed from drips before they fall from ceilings and walls. Straw stalactites are formed when the degassing of single drops of water on the ceiling produces a ring of calcite about 5 mm in diameter. This ring grows downwards as a hollow tube made of a single crystal with its growth axis vertical. The longest such straws recorded are 4.2 m in the Grotte de la Clamouse, France, and 6.2 m in Strong's Cave, Western Australia. Drops falling to the floor build stalagmites, whose growth layers thin away from the point of impact. Constant drip rates, water

Plate 4.5
*Cave coralloid formations
in Resurrection Cave,
Mount Etna, Queensland.*

hardness and cave atmosphere promote uniform diameter stalagmites, while decreasing deposition rates result in conical forms.
Sinuous rills down walls result in curtains and shawls which are but
a single crystal thick. Hill and Forti (1986) provide a very comprehensive review of these and other forms. There is some post-1986
research as well as earlier work not discussed in detail in previous
reviews. For instance:

- moonmilk, a soft powdery calcite, has attracted a lot of
  attention primarily because of its unusual chemical composition and its use by humans. The precise reasons for its
  microcrystalline habit are unclear. Its formation is discussed by Bernasconi (1976, 1981) and in later work by
  Fischer (1988, 1992) and Onac and Ghergari (1993).
- some unusual subaqueous speleothems are noted by Davis
  (1989) and Davis et al. (1990), including helictites draped
  in calcite rafts.
- rare, cubic cave pearls were reported from Castleguard
  Cave by Roberge and Caron (1983). Their flat facets are
  postulated to have formed by contact with neighbouring
  concretions.
- a model for the development of spathite (flaired soda
  straws) was proposed by Hubbard et al. (1984).
- observations on deflected soda straw stalactites in Wales
  by Sevenair (1985) suggested air movements were responsible for non-vertical growth with wind flow directions
  through the cave correlating well with the straw orientation. Davis (1986) suggested the possibility of differential
  seepage deposited minerals having the same effect.

Cave coral or 'popcorn' is a widespread formation (see plate 4.5) whose origin remains unclear. Strictly the formation should be termed 'coralloid' to describe the great variety of nodular, globular, botryoidal and coral-like speleothems subsumed in a single term (Hill and Forti 1986). Clearly some forms have developed under water, while others have formed at the interface of rock and air in both wet and dry environments. The following mechanisms have been suggested by Hill and Forti (1986):

Plate 4.6
Quill anthodites composed of aragonite in Shishkabob Cave, Mole Creek, Tasmania.

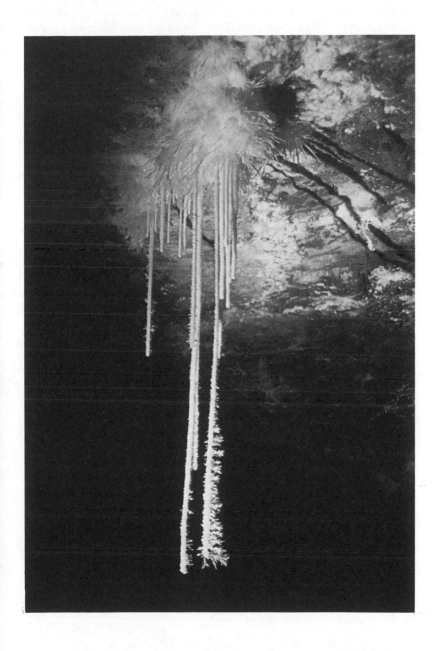

- by bedrock seepage emerging and passing through the coralloid itself
- by thin films of water flowing over wall irregularities
- by splash from dripping water
- by the upwards capillary movement of water from pools onto walls
- by the condensation of water droplets and degassing.

Probably more than one mechanism can be operating at any one site and over time as the cave environment changes. It is still unclear whether organic substrates such as fungal hyphae and plant roots can act as condensation nuclei for coralloid speleothems, though numerous examples of this can be seen.

Colour of calcite formations

Pure calcite and aragonite speleothems are most likely to be translucent and colourless (Ford and Williams 1989). Clastic inclusions, trace elements and organics may impart coloration. White (1981) confirmed Gascoyne's (1977) conclusion that yellow-brown coloration may be caused primarily by high molecular weight and fulvic and humic substances leached from overlying decomposing soils (Lauritzen et al. 1986). Latham (1981) discussed the effect of organic substances on crystal growth.

Important Non-Carbonate Minerals

There is a bewildering array of non-carbonate minerals found in caves. Ford and Williams (1989), quoting Hill and Forti, cite over 180, to which can be added another twenty from the caves of Bulgaria (Shopov 1988). Many of these are unusual phosphatic minerals formed in the presence of bat guano. The most important are evaporites, phosphates and nitrates and oxides and hydroxides.

Evaporites

Evaporitic minerals have their surface analogues in playa/brine deposits of semi-arid to arid regions. Evaporite cave minerals are highly soluble in water and are thus commonly found in dry parts of caves or in caves located within regions receiving less than 250 mm annual average rainfall. They are not necessarily confined to warmer regions, however. Harmon et al. (1983) reported at least three sulphate minerals from Castleguard Cave in the Canadian Rockies (gypsum, mirabilite and epsomite). Powdery and crystalline gypsum deposits are also found deep in caves like Selminum Tem and Atea Kananda in ever-wet New Guinea, where the annual average rainfall exceeds 4000 mm. In these situations the source of the sulphate is pyritic impurities in the limestone or in interbeds, and vigorous air circulation aids evaporation.

Plate 4.7
Gypsum encrustations in
Ozernaja Cave, Ukraine.
Photo by John Gunn.

Cave evaporites consist largely of sulphates and halides. Hill and
Forti (1986) list ten sulphate minerals derived from 'normal' karst
processes; many others result from the interaction of karst waters
with ore bodies. The most common evaporite mineral is gypsum
(see plate 4.7), thought to be the third most common cave mineral
overall after calcite and aragonite (Hill and Forti 1986). Gypsum is
especially abundant in the caves of the American Southwest. Of the
halides, halite is the most abundant form. While the caves of the

Nullarbor Plain, Australia, have some of the largest halite formations known, the caves of Mt Sedom, Israel, are formed entirely in
halite and have extensive halite speleothems.

Gypsum and other sulphate minerals require a source of sulphate
ions. White (1988) lists four possible sources, which can be
recategorized into two distinct source groups:

- Carbonate bedrock sources:
  a) dissolution of sulphate-bearing beds
  b) oxidation of pyrite (FeS_2) occurring in the bedrock
  c) weathering of anhydrite nodules
- Deep groundwater sources:
  a) movement of deep H_2S waters into contact with
     phreatic karst waters to form sulphuric acid, which
     may dissolve calcite and deposit gypsum.

To these, Hill and Forti (1986) add bat guano and basalts.

Together with other evaporites, gypsum deposition via evaporation is often regarded as a physical rather than a chemical process
(Bögli 1980; Bull 1983). However, while gypsum and related sulphate minerals are a result of evaporation, such precipitation is
often a final stage of more complex chemical reactions. This is
illustrated in two of the four mechanisms for gypsum deposition
listed by Hill and Forti (1986). Phase diagrams for sulphate minerals are shown in Hill and Forti (1986).

The sensitivity of sulphates to cave environmental conditions
is illustrated by Maltsev (1990). He found the growth of gypsum
anemolites (wind-controlled speleothems) in a Ukraine cave
varied with cave humidity fluctuations. In summer, when air
flows into the cave and with humidity at 70 to 90 per cent, accretion of up to 7mm occurred. In winter, the air flow is reversed,
humidity rises and dissolution of the fine gypsum structures takes
place.

Halite and other halides are extremely soluble and are confined
to relatively high temperatures and low humidities and, on present
knowledge, are found in only a few caves situated in arid to semi-
arid regions (White 1976). Parent minerals are leached from the
regolith by percolation waters and redeposited under evaporative
conditions in the cave environment. Lowry (1967) discussed halite
speleothems from the Nullarbor; Goede et al. (1992), in describing
from the same region what may be the largest halite speleothems in
the world, hypothesized their forming under conditions of low
humidity and strong air currents.

Exsudation or salt weathering is responsible for the production
of the famous 'coffee and cream' which is perhaps the most dramatic and unusual of the Nullarbor cave sediments. The best and
most extensive deposits were to be found in Mullamullang Cave
but they are also known from a number of other sites. 'Coffee and
cream' is an airfall deposit produced by crystal weathering. It is
made up of a fine crystalline powder and a more granular material

with small fragments of halite and gypsum cave flowers. Both the light- and the dark-coloured materials are made up of a high magnesium calcite. The 'cream' is a cream, occasionally pink, colour containing about 2 per cent of iron minerals. The strikingly contrasting 'coffee' is of similar chemical composition but with about 8 per cent iron compounds plus some manganese dioxide (Caldwell et al. 1982). The two forms appear to flow over one another. The whole forms an unusual entity of puzzling appearance, especially as the roofs above appear to be evenly coloured and textured, at least at a scale far smaller than the 'coffee and cream' facies. They probably deserve further investigation, but unfortunately the Mullamullang Cave deposits have been very much disturbed by visitors. It is believed that 'coffee and cream' is only found in caves beneath the Nullarbor Plain.

Although evaporites form deposits similar to those formed by carbonates (especially stalactites and stalagmites), most halide and sulphate formations are characterized by unique morphologies (see plate 4.8). This is a reflection of both the crystallographic properties of the minerals involved and depositional mechanisms from which they are produced. As most occur from the evaporation of seepage waters, the common forms are crusts, incipient stalactites and stalagmites (e.g., James 1991) and soil cements (e.g., Harmon et al. 1983). Selenite swords are perhaps a spectacular exception – they are thought to be sub-aqueous in origin. Hill and Forti (1986) and White (1976) provide the best descriptions of evaporite speleothems, although the latter is concerned largely with gypsum formations.

Phosphates and nitrates

Phosphate and nitrate minerals are largely organic in origin, which has prompted Bögli (1980) to categorize them as 'organic' cave sediments. According to White (1976, 313), however, 'the reaction products contain authigenic chemical species which must be regarded as cave minerals.' The major source for the phosphates is the waste products and remains of cave-dwelling fauna, especially bats; nitrate sources may derive from within the cave as well as from overlying materials, especially soils (Ford and Williams 1989). Lists of major phosphatic minerals can be found in White (1976), Bögli (1980) and Hill and Forti (1986).

Phosphate and nitrate minerals are formed by the reaction of leached phosphate and nitrate compounds with limestone bedrock, secondary carbonate deposits and/or unconsolidated cave sediments. The phosphate 'solubilization' process, as well as a description of a typical flow path from leachate to the precipitation of various phosphate species, is discussed in Hill and Forti (1986, figure 85). Phosphate minerals rarely occur as pure or semi-pure forms produced by carbonate and evaporite minerals (although see Dunkley and Wigley (1967) for a description of 300 mm

Plate 4.8
Mixed calcite and gypsum
stalagmite in Gua
Tempurung, Malaya.
Photo by Andy Lawrence.

biphosphammite stalactites). They occur primarily as crusts on breakdown material and bedrock walls, or as nodules and horizons within cave clastic sediments or guano piles (Hill and Forti 1986).

Cave nitrate minerals have generated considerable interest because of their use in the production of gunpowder. Hill (1981a) made the distinction between 'saltpetre caves' (i.e., caves containing high nitrate concentrations in their inorganic, clastic sediments) and 'guano caves' (i.e., caves containing nitrous organic bat guano but little clastic sediment). Saltpetre was mined from sediments and, strictly speaking, is a product derived from nitre (KNO_3) and not a primary cave mineral (Hill 1981a).

The formation of cave nitrate minerals is not entirely understood. Hill (1981b) and Hill and Forti (1986) list several nitrate sources: surface soils, volcano-sourced groundwaters and cave guano piles. Since nitrate minerals are formed in the absence of guano piles, it is thought that their origin stems largely from the leaching of nitrates from overlying soils with the assistance from *Nitrobacter* (a nitrogen-fixing bacterium). This has been extensively studied by Hill (1981b). This view has been challenged in recent times by Lewis (1992), who ruled out a seepage water origin and instead argued that bacteria in cave sediments fix the nitrogen from the cave atmosphere.

Because of their high solubility, nitrates are confined to very dry sections of caves and to caves in very dry regions. Hill (1981b) provides a phase diagram for the major nitrate minerals from US caves. They form mainly crust deposits on cave walls as well as coatings on guano piles and on and within cave soils (Hill and Forti 1986).

Oxides, silicates and hydroxides

While many oxides, silicates and hydroxides may be transported into the cave environment, some are chemically formed *in situ*. The cave environment is wet, mildly alkaline and oxidizing, conditions which favour the formation of certain Fe and Mn oxide minerals (White 1976). The source ions for these Fe and Mn deposits are weathered surficial deposits and decaying organic matter. White (1976) and Hill and Forti (1986) provide the most useful discussions.

Manganese deposits most commonly occur as dark coatings on stream bottoms and cave walls (Ford and Williams 1989) but may form unconsolidated fills (Hill and Forti 1986). Deposition is a result of mixing of Mn-bearing water with highly oxygenated cave water; bacteria may be involved in their precipitation (Moore 1981; Peck 1986). Hill (1982) has documented these and other black deposits from US caves. Birnessite is thought to be the most common Mn oxide (Moore 1981); others include pyrolusite and ranciéite (Laverty and Crabtree 1978), the former being common in surface soils and in calcretes.

Iron oxide minerals form from the oxidation of pyrite or other iron-bearing sources (White et al. 1985; White 1988); reduction by bacteria may also be important (Peck 1986; Hill and Forti 1986; Davis et al. 1990). Haematite usually indicates thermal conditions (Hill and Forti 1986). Unlike Mn oxides, Fe minerals may form more extensive deposits, including stalactites and stalagmites (up to 3 m high – Szczerban and Urbani 1974).

The two most common forms of siliceous minerals are quartz and cristobolite (White 1976). Hill and Forti (1986) cite a range of silicate minerals (especially clays) whose origin may be detrital rather than secondary; other silicate minerals, such as allophane (Webb and Finlayson 1984, 1985), are secondary. Most silica is derived from the weathering of non-carbonate rocks, particularly igneous. Thus, cave groundwaters previously in contact with such sources may be rich in dissolved silica. Precipitation via evaporation and CO_2 degassing are the most common depositional processes (Hill and Forti 1986). White (1976) suggested that large euhedral quartz forms may indicate elevated temperatures of the depositing waters. Siliceous deposits occur in forms similar to those produced by carbonates (Hill and Forti 1986).

Ice speleothems are found in alpine and/or high latitude caves or in caves characterized by a particular set of air-flow conditions capable of maintaining low cave temperatures (White 1976). Ice can form either seasonally (usually as winter ice close to cave entrances) or it can be permanent (Ford and Williams 1989). Caves supporting perennial ice are known as 'glacieres'. Cave ice forms from two mechanisms: the freezing of seepage water and the freezing of water vapour. According to Ford and Williams (1989), these produce at least seven types of cave ice: dripstones and flowstones; recrystallized 'old' ice; frozen ponds; hoarfrost (of which four forms are recognized); extrusion ice; intrusion ice; and as an ice matrix in clastic sediments. Detailed descriptions of ice formations are provided in Hill and Forti (1986).

Other Minerals

Hill and Forti (1986) categorize other minerals under 'ore-related' and 'miscellaneous'. Ore-related minerals are especially interesting; they precipitate from groundwaters in contact with igneous bodies and may produce in any one cave a varied mineral assemblage. An excellent example is from Lilburn Cave, California (Rogers and Williams 1982), where petromorphic minerals, such as hornblende, occur in abundance.

Cave Formations of the Nullarbor Plain, Australia

The Nullarbor Plain is Australia's largest karst ($220\,000\,km^2$), with an arid to semi-arid climate: annual rainfall is between 150 and 250 mm while annual evaporation is 1250 to 2500 mm. The plain has no surface drainage, but is punctuated by collapse dolines which are generally shallow. The caves are extensive, much modified by salt wedging and collapse processes, and commonly descend gently to near-static pools and lakes of brackish to saline water. The flooded tunnels of Cocklebiddy Cave are in excess of 6 km long – most of them explorable only by scuba diving. Withdrawal of hydrostatic support at times of lower sea level has led to collapse into these water table caves, allowing entry into large caverns such

as Koonalda, Abrakurrie (see plate 4.9) and Weebubbie caves (see figure 4.2).

A particular feature of the Nullarbor caves is the abundance of halite, which produces both speleothems (see plate 4.10) and many weathering forms through wedging during recrystallization. Gypsum speleothems are also abundant, with calcite speleothems less obvious largely because of destruction through salt wedging. Lowry and Jennings (1974) provide a good introduction to the geomorphology, while more recent detail is given in the review by Gillieson and Spate (1992).

The Nullarbor caves contain secondary minerals of great significance, a number of which have been identified as new minerals. More common minerals exist in previously unrecorded forms and many of the mineral deposits are of great value as a source of palaeoenvironmental data for the Nullarbor Plain. Table 4.3 lists the secondary minerals that have been identified from the Nullarbor caves, their chemical formula and location. Key references are Bridge (1973), Bridge and Clarke (1983) and Caldwell et al. (1982).

The Nullarbor contains one speleothem form, closely allied to cave shields, which appears to be different from those reported

Plate 4.9
The main passage of Abrakurrie Cave, Nullarbor Plain, is typical of chambers that have been highly modified by exsudation. The floor area of the chamber is 9000 m².

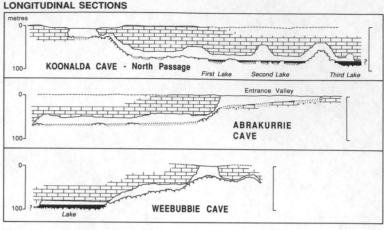

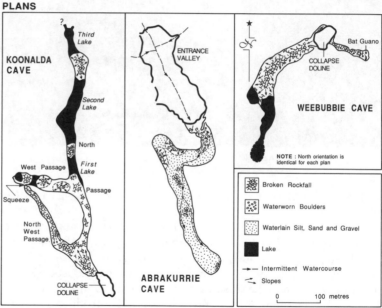

Figure 4.2
Longitudinal sections and plans of some deep Nullarbor Plain caves. From Gillieson and Spate (1992).

elsewhere. The term 'stegamite' was coined by Webb (1991), who described it as a high ridge in a calcite floor, generally with a crack along its top, which separates the speleothem into two vertically standing plates. Stegamites are made up of finely banded black calcite from the medial crack in a similar fashion to that postulated for cave shields (Hill and Forti 1986), although there is some dispute as to their origin. They are sometimes found in clusters, and in a few cases secondary stegamites intersect the primary form approximately at right angles.

The Nullarbor is unique among world karsts in demonstrating very extensive modification of caves by the arid zone process of

Plate 4.10
Halite speleothems about 30 cm long in the Easter Extension, Mullamullang Cave, Nullarbor Plain.

exsudation (otherwise known as salt or crystal weathering). The process works by detachment of particles of various sizes from a rock surface (see plate 4.11) by the growth of crystals (mainly of gypsum, to a lesser extent halite) from percolating saline solutions. This process enlarges hollows in cave roofs to form domes, some of large size: in Mullamullang Cave there are several which are tens of metres in diameter, arching up from the general collapse surface of the original roof by several metres. This doming process may

Table 4.3 Secondary minerals in the Nullarbor caves (from Gillieson and Spate 1992)

Mineral	Composition	Location
Halides		
Halite	$NaCl$	Most caves
Sulphates		
Gypsum	$CaSO_4 \cdot 2H_2O$	Most caves
Selenite	$CaSO_4 \cdot 2H_2O$	Mullamullang N37, Thampanna N206
Mirabilite	$Na_2SO_4 \cdot 10H_2O$	Unknown cave
Taylorite	$(NH_4,K)SO_4$	Murra-el-Elevyn N47
Aphthitalite	$(K,Na)_3Na(SO_4)_2$	Petrogale N200, Murra-el-Elevyn N47
Syngenite	$K_2Ca(SO_4)_2 \cdot 2H_2O$	Murra-el-Elevyn N47
Carbonates		
Calcite	$CaCO_3$	Most caves
Aragonite	$CaCO_3$	Tommy Grahams N56
Phosphates		
Whitlockite	$Ca_9(Mg,Fe)H(PO_4)_7$	Petrogale N200, Murra-el-Elevyn N47
Stercorite	$H(NH_4)Na(PO_4) \cdot 4H_2O$	Petrogale N200
Newberyite	$MgHPO_4 \cdot 3H_2O$	Petrogale N200
Mundrabillaite	$(NH_4)_2Ca(HPO_4)_2 \cdot 2H_2O$	Petrogale N200
Monetite	$CaHPO_4$	Murra-el-Elevyn N47
Hannayite	$(NH_4)_2Ca(HPO_4)_2H_2O$	Murra-el-Elevyn N47
Carbonate-hydroxylapatite	$Ca_5(PO_4,CO_3)_3OH$	Murra-el-Elevyn N47
Brushite	$CaHPO_4 \cdot 2H_2O$	Murra-el-Elevyn N47
Biphosphammite	$(NH_4,K)H_2PO_4$	Petrogale N200, Murra-el-Elevyn N47
Archerite	$(K,NH_4)HPO_4$	Petrogale N200
Organics		
Guanine	$C_5H_3(NH_2)N_4O$	Murra-el-Elevyn N47
Ocammite	$(NH_4)_2C_2 \cdot 4H_2O$	Petrogale N200
Uricite	$C_5H_4N_4O_3$	Dingo Donga N160
Weddellite	$CaC_2O_4 \cdot 2H_2o$	Webbs N132, Petrogale N200
Iron and manganese oxides		
Goethite	$FeOOH$	Mullamullang N37
Haematite	Fe_2O_3	Mullamullang N37
Pyrolusite	MnO_2	Mullamullang N37
Clays		
Illite, Kaolin, Montmorillonite		From caves in the Mundrabilla area

extend vertically upwards from cave voids below to create the vertical shafts which intersect the ground surface as blowholes. Much of the speleothem decoration (particularly the calcite) has been broken down by exsudation.

Another fine textured product of this crystal weathering process is the banks of low density white and brown streaked detritus known as 'coffee and cream'. Exsudation is also capable of shattering massive calcite speleothems. The crystal weathering splits massive speleothems as if they had been sawn down the middle, or smashes them up as if a determined vandal had taken to them with a sledge hammer. Outstanding examples occur in the central Nullarbor, particularly in Webbs Cave, Kelly Cave, Witches Cave and Thampanna Cave.

Plate 4.11
Salt exsudation promoting spalling of the limestone walls in Mullamullang Cave, Nullarbor Plain.

References

Atkinson, T. C. 1983: Growth mechanisms of speleothems in Castleguard Cave, Columbia Icefields, Alberta, Canada. *Arctic and Alpine Research* 15, 523–36.

Bernasconi, R. 1976: The physico-chemical evolution of moonmilk. *Cave Geology* 1, 63–88.

Bernasconi, R. 1981: Mondmilch (moonmilk): two questions of terminology. *Proc. 8th Int. Congr. Speleo., Kentucky*, 113–16.

Bischoff, J. L. and Fyfe, W. S. 1968: Catalysis, inhibition and the calcite–aragonite problems, I: the aragonite–calcite transformation. *American Journal of Science* 266, 65–79.

Bögli, A. 1980: *Karst Hydrology and Physical Speleology*. Berlin: Springer, 284 pp.

Bridge, P. J. 1973: Guano minerals from Murra-el-Elevyn Cave, Western Australia. *Mineralogical Magazine* 39, 467–9.

Bridge, P. J. and Clarke, R. M. 1983: Mundrabillaite: a new cave mineral from Western Australia. *Mineralogical Magazine* 47, 80–1.

Bull, P. A. 1983: Chemical sedimentation in caves. In A. S. Goudie and K. Pye (ed.) *Chemical Sediments and Geomorphology: Precipitates and Residua in the Near-Surface Environment.* London: Academic Press, 439 pp.

Cabrol, P. and Coudray, J. 1978: Les phénomènes de diagenèse dans les concrétions carbonatées des grottes. *6ème Réunion Annuelle des Sciences de la Terre.* Orsay, France, 82–5.

Caldwell, J. R., Davey, A. G., Jennings, J. N. and Spate, A. P. 1982: Colour in some Nullarbor Plain speleothems. *Helictite* 20 (1), 3–10.

Craig, K. D., Horton, P. D. and Reams, M. W. 1984: Clastic and solutional boundaries as nucleation surfaces for aragonite in speleothems. *NSS Bull.* 46, 15–17.

Curl, R. L. 1962: The calcite–aragonite problem. *NSS Bull.* 24, 57–73.

Curl, R. L. 1972: Minimum diameter stalactites. *NSS Bull.* 34, 129–36.

Curl, R. L. 1973: Minimum diameter stalagmites. *NSS Bull.* 35, 1–9.

Davis, D. G. 1986: Discussion: the deflected stalactites of Dan-Yr-Ogof. *NSS Bull.* 48, 62–3.

Davis, D. G. 1989: Helictite bushes – a subaqueous speleothem? *NSS Bull.* 51, 120–4.

Davis, D. G., Palmer, A. N. and Palmer, M. J. 1990: Extraordinary subaqueous speleothems in Lechuguilla Cave, New Mexico. *NSS Bull.* 52, 70–86.

Davis, D. G., Palmer, A. N. and Palmer, M. V. 1992: Extraordinary subaqueous speleothems in Lechuguilla Cave, New Mexico: reply. *NSS Bull.* 54, 34–5.

Dreybrodt, W. 1980: Deposition of calcite from thin films of natural calcareous solutions and the growth of speleothems. *Chemical Geology* 29, 89–105.

Dreybrodt, W. 1982: A possible mechanism for growth of calcite speleothems without participation of biogenic carbon dioxide. *Earth and Planetary Science Letters* 58, 293–9.

Dunkley, J. R. and Wigley, T. M. L. 1967: *Caves of the Nullarbor.* University of Sydney: Speleological Research Council, 61 pp.

Fischer, H. 1988: Etymology, terminology and an attempt of definition of Mondmilch. *NSS Bull.* 50, 54–8.

Fischer, H. 1992: Type locality of Mondmilch. *Cave Science* 19, 59–60.

Folk, R. L. and Asserto, R. 1976: Comparative fabrics of length-slow and length-fast calcite and calcitized aragonite in a Holocene speleothem, Carlsbad Cavern, New Mexico. *Journal of Sedimentary Petrology* 46, 486–96.

Ford, D. C. and Williams, P. W. 1989: *Karst Geomorphology and Hydrology.* London: Unwin Hyman, 601 pp.

Franke, H. W. 1975: Sub-minimum diameter stalagmites. *NSS Bull.* 37, 17–18.

Gams, I. 1981: Contribution to morphometrics of stalagmites. *Proc. 8th Int. Congr. Speleo., Kentucky,* 276–8.

Gascoyne, M. 1977: Trace element geochemistry of speleothems. *Proc. 7th Int. Congr. Speleo., Sheffield,* 205–8.

Gillieson, D. S. and Spate, A. P. 1992: The Nullarbor karst. In D. S. Gillieson (ed.) *Geology, Climate, Hydrology and Karst Formation: Field*

Symposium in Australia: Guidebook. Special Publication No. 4, Department of Geography and Oceanography, University College, Australian Defence Force Academy, Canberra, 65–99.

Goede, A., Atkinson, T. C. and Rowe, P. J. 1992: A giant Late Pleistocene halite speleothem from Webbs Cave, Nullarbor Plain, southeastern Western Australia. *Helictite* 30, 3–7.

Gonzalez, L. A. and Lohmann, K. C. 1988: Controls on mineralogy and composition of spelean carbonates: Carlsbad Caverns, New Mexico. In N. P. James and P. V. Choquette (ed.) *Paleokarst*. New York: Springer, 416 pp.

Harmon, R. S., Atkinson, T. C. and Atkinson, J. L. 1983: The mineralogy of the Castleguard Cave, Columbia Icefields, Alberta, Canada. *Arctic and Alpine Research* 15, 503–16.

Hill, C. A. 1976: *Cave Minerals*. Huntsville, AL: Nat. Speleo. Soc.

Hill, C. A. 1981a: Origin of cave saltpeter. *NSS Bull.* 43, 110–26.

Hill, C. A. 1981b: Mineralogy of cave nitrates. *NSS Bull.* 43, 127–32.

Hill, C. A. 1982: Origin of black deposits in caves. *NSS Bull.* 44, 15–19.

Hill, C. A. and Forti, P. 1986: *Cave Minerals of the World*. Huntsville, AL: Nat. Speleo. Soc.

Hubbard, D. A., Herman, J. S. and Mitchell, R. S. 1984: A spathite occurrence in Virginia: observations and a hypothesis for genesis. *NSS Bull.* 46, 10–14.

James, J. M. 1977: Carbon dioxide in the cave atmosphere. *Trans. Brit. Cave Res. Assoc.* 4, 417–29.

James, J. M. 1991: The sulphate speleothems of Thampanna Cave, Nullarbor Plain. *Helictite* 30, 19–23.

Jennings, J. N. 1985: *Karst Geomorphology*. Oxford: Blackwell.

Kendall, A. C. and Broughton, P. L. 1978: Origin of fabrics in speleothems composed of columnar calcite crystals. *Journal of Sedimentary Petrology* 48, 519–38.

Latham, A. G. 1981: Muck spreading on speleothems. *Proc. 8th Int. Congr. Speleo., Kentucky*, 356–7.

Lauritzen, S. E., Ford, D. C. and Schwarcz, H. P. 1986: Humic substances in speleothem matrix: Paleoclimatic significance. *Proc. 9th Int. Congr. Speleo., Barcelona*, 77–9.

Laverty, M. and Crabtree, S. 1978: Ranciéite and mirabilite: some preliminary results on cave mineralogy. *Trans. Brit. Cave Res. Assoc.* 5, 135–42.

Lewis, W. C. 1992: On the origin of cave saltpeter: a second opinion. *NSS Bull.* 54, 28–30.

Lowry, D. C. 1967: Halite speleothems from the Nullarbor Plain, Western Australia. *Helictite* 6, 14–20.

Lowry, D. C. and Jennings, J. N. 1974: The Nullarbor karst, Australia. *Zeitschrift für Geomorphologie* 18, 35–81.

Maltsev, V. A. 1990: The influence of seasonal changes of cave microclimate upon the genesis of gypsum formations in caves. *NSS Bull.* 52, 99–103.

Moore, G. W. 1956: Aragonite speleothems as indicators of palaeotemperature. *American Journal of Science* 254, 746–53.

Moore, G. W. 1962: The growth of stalactites. *NSS Bull.* 24, 95–106.

Moore, G. W. 1981: Manganese deposition in limestone caves. *Proc. 8th Int. Congr. Speleo., Kentucky*, 642–4.

Onac, B.-P. and Ghergari, L. 1993: Moonmilk mineralogy in some Romanian and Norwegian caves. *Cave Science* 20, 107–11.

Peck, S. B. 1986: Bacterial deposition of iron and manganese oxides in North American caves. *NSS Bull.* 48, 26–30.

Picknett, R. G., Bray, L. G. and Stenner, R. D. 1976: The chemistry of cave waters. In T. D. Ford and C. H. D. Cullingford (ed.) *The Science of Speleology.* London: Academic Press, 593 pp.

Roberge, J. and Caron, D. 1983: The occurrence of an unusual type of pisolite: the cubic cave pearls of Castleguard Cave, Columbia Icefields, Alberta, Canada. *Arctic and Alpine Research* 15, 517–22.

Rogers, B. W. and Williams, K. M. 1982: Mineralogy of Lilburn Cave, Kings Canyon National Park, California. *NSS Bull.* 44, 23–31.

Roques, H. 1964: Contribution à l'étude statique et cinétique des systèmes gaz carbonique-eaucarbonate. *Ann. Spéléo.* 19, 255–484.

Sevenair, J. P. 1985: The deflected stalactites of Dan-Yr-Ogof. *NSS Bull.* 47, 28–31.

Shopov, Y. Y. 1988: Bulgarian cave minerals. *NSS Bull.* 50, 21–4.

Szczerban, E. and Urbani, F. 1974: Carsos de Venezuela, 4: Formas carsicos en areniscas precambricas del Territorio Federal Amazonas y Estado Bolivar. *Bol. Soc. Venezolana Ispel.* 5 (1), 27–54.

Thrailkill, J. V. 1968b: Dolomite cave deposits from Carlsbad Caverns. *Journal of Sedimentary Petrology* 38, 141–5.

Varnedoe, W. W. 1965: A hypothesis for the formation of rimstone dams and gours. *NSS Bull.* 27, 151–2.

Webb, J. A. and Finlayson, B. L. 1984: Allophane and opal speleothems from granite caves in south-east Queensland. *Aust. J. Earth Sci.* 31, 341–9.

Webb, J. A. and Finlayson, B. L. 1985: Allophane flowstone from Newton Cave, western Washington State. *NSS Bull.* 47, 45–8.

Webb, R. (1991): Stegamites – a form of cave shield? *Proc. 18th Bienn. Conf. Aust. Speleo. Fedn,* 95–8.

White, W. B. 1976: Cave minerals and speleothems. In T. D. Ford and C. H. D. Cullingford (ed.) *The Science of Speleology.* London: Academic Press, 593 pp.

White, W. B. 1981: Reflectance spectra and color in speleothems. *NSS Bull.* 43, 20–6.

White, W. B. 1982: Mineralogy of the Butler Cave–Sinking Stream system. *NSS Bull.* 44, 90–7.

White, W. B. 1988: *Geomorphology and Hydrology of Karst Terrains.* New York: Oxford University Press, 464 pp.

White, W. B., Scheetz, B. E., Atkinson, S. D., Ibberson, D. and Chess, C. A. 1985: Mineralogy of Rohrer's Cave, Lancaster County, Pennsylvania. *NSS Bull.* 47, 17–27.

Zeller, E. J. and Wray, J. L. 1956: Factors influencing the precipitation of calcium carbonate. *Bull. Am. Soc. Petrol. Geol.* 40, 140–52.

FIVE

Cave Sediments

Introduction

Early in their underground career most cavers will become intimately acquainted with the unctuous cave muds which adhere to just about anything that contacts them. They will also be aware of the seemingly vast amount of haphazardly piled angular boulders which make up cave breakdown. The reward for negotiating these breakdown piles and mud wallows is the privilege of viewing pristine calcite formations. When most people think of cave sediments, they have in mind the clastic sediments made up of fine or coarse particles of mineral or organic matter. A great deal of research has been carried out on the material deposited with these clastic sediments (bones, pollen, artefacts) as a means of elucidating environmental or human histories. Less research has been undertaken on the processes by which clastic sediments are produced, transported and deposited within the cave system. In part this is owing to the difficulty of observing and measuring these processes in flooding caves. As we have seen in Chapter 1, the bulk of material moving through a cave system is clastic sediment. There is in fact very little difference in the nature of surface and underground clastic sediments, although we must devise more time-transgressive models of sedimentation for cave sediments. In addition, there is an important interaction between cave sediments and cave morphology which produces a range of depositional structures and passage morphologies. Cave sediments are tricky to understand!

Clastic Sediment Types

Clastic sediments form from fragments of rocks (the regolith) which have been broken up by physical or chemical weathering processes. These fragments are further transformed by the winnowing effect of sediment transport and by chemical alteration

Table 5.1   Cave sediment types

Type	Origin	Nature
Clastic	Allogenic or authigenic	Angular boulder debris; subangular to subrounded gravels and cobbles; sands and silts; cave clays
Organic	Allogenic debris	Woody debris; humus and fine particulate organic matter; dung and spores
	Authigenic deposits	Bat and bird guano; phosphatic mineral crusts and laminae; nitrate-rich deposits; subaerial stromatolites
Chemical (see Chapter 4)	Allogenic	Tufa and travertine fragments; pisolites (iron nodules); caliche
	Authigenic	Calcite and gypsum speleothems and interlayered deposits; tufas and travertines
Ice	Authigenic cave drips, wall condensation, cave pools	Ice stalagmites and stalactites; pond ice; rime crystals

(diagenesis) during long or short periods of repose in the surface or underground environment. Not only the geologic origin of the sediments, but also their place of origin, determines their classification as either allogenic (origin outside the cave) or authigenic (origin within the cave). Cave sediment types (see table 5.1) are thus diverse and include organic debris and its chemical derivatives (phosphate and nitrate minerals), inorganic chemical precipitates, and ice. In any single cave the total assemblage will depend on both the past and present geologic and climatic environment. Cave sediments thus provide us with a potential library of environmental information – if we can read the language in which they are written. To do this we need to understand the nature of the materials and their processes of transport, deposition and diagenesis.

Processes of Sedimentation

Cave sediments may be deposited by either gravity-fall or aqueous transport processes. The distinction between these becomes blurred when we consider such processes as turbidity currents sliding down steep sediment banks in a cave pool, or the injection of fluidized mudflows into tropical cave passages by mass movements (Gillieson 1986).

Gravity-fall processes

These involve the slow or rapid movement of clastic sediments in air, either as a dry or as a saturated mass movement. The deposition of air-fall tephra (volcanic ash) in caves is a special case of this type. The principal types of deposits moved by gravity-fall processes include cave breakdown, debris cones under avens, cave wall fans below fissures, loess and related deposits, bushfire smoke and fine charcoal, mudflow deposits and till. In addition the deposition of lint and skin cells in tourist caves is a major process of sedimentation by gravity fall, and may total several tonnes per year in heavily used caves.

By far the most widespread form of gravity-fall sediment in caves is breakdown or incasion. The alteration of passage shape by breakdown seems to be the ultimate fate of most caves once the water which formed them goes elsewhere. While the passage is water filled, stress lines in the rock are evenly distributed around the cavity (see figure 5.1); once the water is removed, local concentration of stress leads to failure of the arched section, usually along bedding planes. This process propagates upwards, leading to the development of breakdown domes or avens. In many cases this propagation intersects overlying non-limestone rocks, allowing this allogenic material to enter the cave.

Breakdown blocks vary in dimension from fist-sized rocks to boulders the size of houses. Their shape depends greatly on the thickness of the bedding. Thus a thinly bedded limestone will tend to produce platy fragments, while a massive limestone will tend to produce more cubic fragments. This purely stress-related breakdown may be modified by frost shattering near entrances to produce platy, angular debris of cobble to boulder size. Thus in Greftsprekka Cave, Nordland, Norway, frost-shattered debris can be found up to 50 m below the surface (see plate 5.1), while more massive breakdown is found throughout the cave to a depth of 230 m.

The stages in development of a passage by breakdown (see plate 5.2) are illustrated from Postojna Cave, Slovenia, by Gospodaric (1976). Here cave development by the underground Pivka River has persisted throughout several glacial–interglacial cycles (see figure 5.2). At the earliest stages, in the mid-Quaternary, the fluvial regime is predominantly erosional with little active sedimentation. Allogenic chert gravels and sands eroded the cave passage, and at a later stage quartz sands, chert sands and limonite from bauxitic soils were deposited. An initial breakdown cone formed in the active stream passage; its development ponded water and led to the excavation of a lower passage. The first speleothems formed in the passage, principally stalagmites and reddish flowstones. During the third stage the Pivka eroded and transported rubble, gravel and flood loam. Collapse of the speleothems occurred owing to sapping of the sediments from below and removal of the collapsed material in the lower course of the Pivka. In the most recent phases of development the breakdown dome has stoped to intersect the surface and form the Stara Apnenica doline. Thus there is an alternation of fluvial and breakdown processes, with the break-

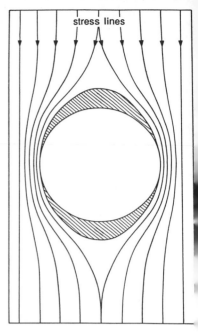

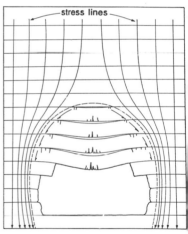

Figure 5.1
Distribution of stress lines around natural cavities in limestone: (a) distribution around water-filled void below water table; (b) distribution around air-filled void after lowering of water table. Note incipient collapses at areas of concentrated stress.

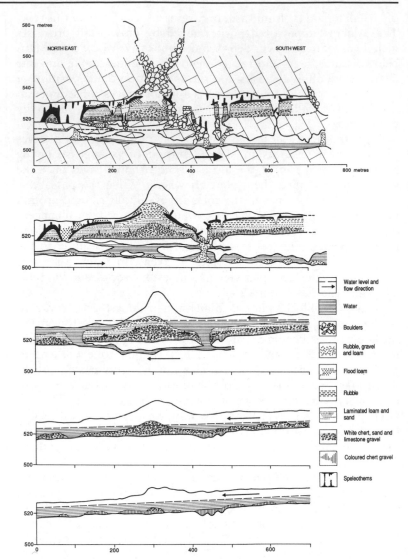

Figure 5.2
Developmental phases of the Otoska Jama, a part of the Postojna caves of Slovenia. The top diagram shows the present cave morphology in cross-section, dominated by collapse. The bottom diagram is the inferred original morphology of the phreatic passages excavated by the Pivka River. Progressive lowering of the level of the Pivka has led to several phases of cave roof collapse. Modified from Gospodaric (1976).

down sealing and therefore protecting the underlying waterlain sediments. Extensive speleothem deposition has occurred in the dry galleries abandoned by the underground Pivka River. Thus the contemporary distribution of sediments in the cave reflects a long history of deposition and erosion by a variety of processes. The elucidation of this sequence relies on careful inspection, analysis and correlation of sediment sequences throughout the cave. The sequence presented above may not be visible at any one point in the cave; it is an abstraction subject to revision in the light of future discoveries and refinements in technique. Thus while initial age estimates based on radiocarbon suggested a minimum age of c.40 000 years BP for the flood loams in Postojna Jama, later

Plate 5.1
The 50m entrance shaft of Greftsprekka Cave, Norway, has been heavily modified by frost shattering during colder climatic phases.

electron spin resonance studies (Ikeya et al. 1983) indicated that these deposits may be older than 190 000 years BP. This initial underestimation of the age of cave sediments seems to be generally the case, with examples coming from Australia (Frank and Jennings 1978) and Yorkshire (Gascoyne et al. 1983).

Waterlain clastic sediments

Caves can be seen as underground gorges and floodplains, in which clastic sedimentation proceeds in modes analogous to those of

Plate 5.2
*Collapse chamber near the
entrance of Ogof Ffynnon
Dhu, South Wales. Photo
by Gareth Davies.*

surface fluvial systems. This provides a conceptual scheme for
sedimentation processes where water is the transporting agent. The
major difference between the surface and underground fluvial systems is that, in the latter, the water and sediment is confined within
a conduit. This results in two main effects:

- Dramatic fluctuations in water level owing either to flood
  stage or to passage morphology result in steep gradients in
  depositional energy along a cave passage. There is thus
  greater diversity of sediment textures per unit length of
  channel than on the surface. This affects both estimation
  of past flow velocities from sediments and stratigraphic
  correlation.
- Subsequent flows of water down a particular cave passage
  may wholly or partially remove the sediment deposited by
  a prior event. The resistance of an individual 'parcel' of
  sediment to this process of reworking will depend on its
  texture and on the passage geometry at the site. The fate of
  a parcel of sediment entering a cave is to be shunted
  through the passage with successive removals of some
  constituents by sorting, and a gradual diminution of total
  volume (see figure 5.3). Thus the life history of a parcel of
  cave sediment is one of periodic reworking until its iden-

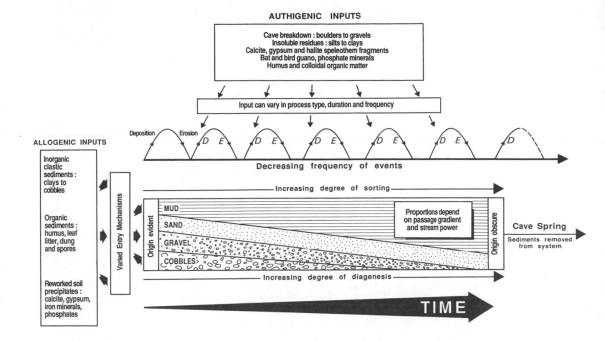

tity is lost, its volume becomes negligible or it is placed in a very low energy environment.

The degree of this reworking will depend largely on the texture of the sediment. Thus the large particles of boulder size and the very fine cohesive clays will both be resistant to reworking once emplaced in a cave. In Westmoreland Cave, Mole Creek, Tasmania, dolerite boulders of Last Glacial age are wedged in the passages and appear little altered by weathering. In contrast, sand-sized sediments will be readily moved and reworked. This is a result of the velocity of erosion being higher than the velocity of transport for coarse and extremely fine particles. In Mammoth Cave, Kentucky, backflooding deposits of silt may be moved to higher elevations by successive floods (Collier and Flint 1964). After every flood, a thin layer of clay is deposited over all submerged passages. This phenomenon is very common in epiphreatic caves, and may act to retard solutional attack on cave walls in the tropics (Gillieson 1985).

Some parcels of sediment may be shunted into side passages during very high floods and will there remain unaltered beyond the reach of successive flow events. Other parcels may be sealed in by rapid flowstone growth or capped by very fine muds which are resistant to erosion. Thus the Cricket Muds of Clearwater Cave, Sarawak, have sat undisturbed (apart from the burrowing of crickets) for the last three-quarters of a million years (Noel and Bull 1982). These very fine-grained muds often have a mean grain size

Figure 5.3
Processes affecting cave sediments through time. Cave sediments may be derived from allogenic or authigenic sources, and may be highly variable in texture. They are sorted by phases of underground transport, which progressively removes some components and leads to an overall reduction in volume. The residuum may eventually emerge at cave springs.

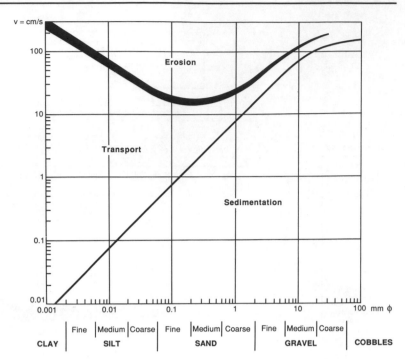

Figure 5.4
Empirical relationships between mean particle diameter in mm and stream velocity in cm s⁻¹ according to Hjulström (1935). After Bögli (1980).

finer than 0.001 mm (10 phi). They are difficult to entrain, but are easily transported as suspended load according to the Hjulström curve (see figure 5.4). On settling from suspension they may accrete at steep angles, draped on underlying rock, flowstone or sediment surfaces (Bull 1981). Quite commonly minor slumping occurs, producing flame structures analogous to those in turbidites, of which they are really a specialized case. Bull (1981) has described surge marks in these fine muds produced by the swash and backwash of pulsed floods in cave passages. These fine muds are useful in that they may provide a palaeomagnetic record and may also preserve pollen. In Castleguard Cave, Aberta, Canada, there are three phases of fine silt filling which are interpreted by Schroeder and Ford (1983) as varved sequences deposited under full glacial conditions. These three silts persist for several kilometres in the cave and were clearly deposited under passage-full conditions. The youngest silt is older than flowstone dated at c.140000 years BP, but is younger than a stalagmite which is >720000 years BP. The other two silts are believed by Schroeder and Ford (1983) to be younger than this stalagmite, but Gale (pers. comm) has isolated pollen from the silts which suggests lowland warm-temperate rainforest of probable Late Miocene age. This material may be reworked from older terrestrial deposits. Thus the precise age of the extensive silt fill of Castleguard Cave is open to conjecture. It is clear from studies in tropical caves (Smart et al. 1986; Williams et al. 1986) that these fine cave fills may be very old and may predate the existing terrain by several million years.

Extensive gravel trains in cave passages may relate to the winnowing of fines by successive flows of water which are incompetent to move the gravels themselves (though some slight movement by bedload traction may occur). The shape of the gravels are held to be different in caves: Siffre and Siffre (1961) suggested that they were flatter because of their transport through inverted siphons, though Bull (1978) thought that flattening was more likely related to thin bedding. More recently Kranjc (1981) has shown a marked decline in gravel roundness with distance of within-cave transport. Other downstream trends are indicated by Valen and Lauritzen (1990) for erratic boulders in Sirijorda Cave, Norway. There is a consistent decrease in mean grain size away from the points of injection of the boulders, presumably transported in subglacial meltwaters. The boulders are up to one metre in diameter, and extremely large discharges must have been involved.

In Postojna Cave, Slovenia, examination of a sediment section in the underground River Pivka allows reconstruction of a partial history of the cave passage. A false floor of flowstone indicates that the sediment fill once nearly sealed the passage (see figure 5.5). While it was intact a flood loam was deposited on top. Since that time incision has exposed the underlying sediments, interlayered gravels and loams. It is clear that the flow which eroded the loam (sample 404) deposited the gravel (sample 405), as they interfinger in both cross-section and long-section. The whole deposit has a downflow dip, which suggests that it infilled a pool, much like a surface point bar deposit. Consideration of the cumulative particle size curves (see figure 5.5) reveals that the lowest gravels are poorly sorted (wide range of sizes and large dispersion about the mean grain size), and that sorting improves up the section, although mean particle size increases. This suggests increased fluvial energy prior to the deposition of the loam. An increase in depositional energy also occurred after the loam was laid down. Clearly flow in the passage declined or stopped so that the speleothems could be formed. The mineralogy of the gravels and sands suggests that each size component of the sediment has a distinctive source in the catchment, with the coarser gravels being dominantly limestone while the sands are quartz rich. By dating the speleothems it would be possible to gain a minimum age for the gravels and a maximum age for the flood loams; this would allow inferences about climatic control on the sedimentation to be tested.

Careful analysis of individual sediment sections (see plate 5.3) may permit estimation of both the regime and energy of past water flows through a cave passage. This estimation is based on hydraulic and sediment transport theory developed for open channels on the surface (Allen 1985), so results must be interpreted with caution. Nevertheless some useful reconstructions have been made. In the Planinska Jama (Slovenia), Gospodaric (1976) has used the

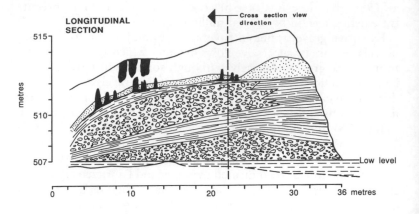

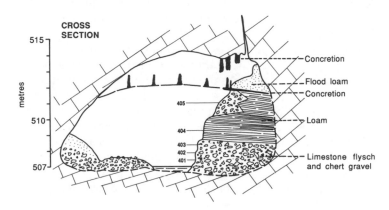

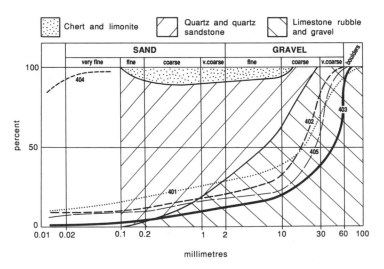

Figure 5.5
Stratigraphy and particle-size characteristics of cave sediments from Postojna Cave, Slovenia. These sediments have been transported by the underground Pivka River and display many stratigraphic characteristics of fluvial sediments, sealed by a flowstone layer. The lower diagram provides details of the particle-size characteristics of sediment samples whose locations are given in the middle cross-section. After Gospodaric (1976).

Plate 5.3
A flowstone dated by uranium series methods at c.50000 BP caps deep fluvial gravels in Eagles' Nest Cave, Yarrangobilly, New South Wales.

particle-size characteristics of fluvial sediments deposited from the River Rak to reconstruct water velocities (see figure 5.6). In the Great Hall an eroded fill of laminated loam is overlain by gravels, the whole section being partially buried by breakdown blocks. Using hydraulic theory the velocity of deposition can be shown to have been initially low, then increased during the phase of gravel deposition. There is a trend to diminishing velocity with the younger sediments, expectable with declining flows through a passage. The sediment textures themselves show a fining upwards trend from gravels through coarse and medium sands. While most

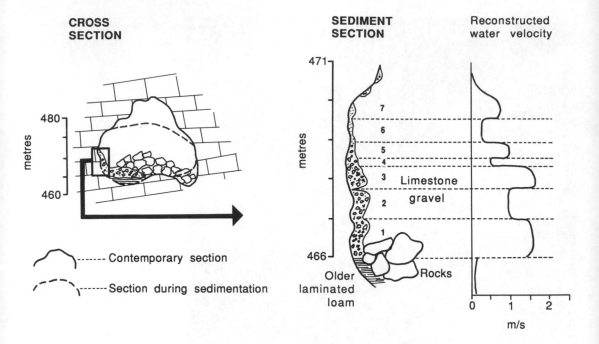

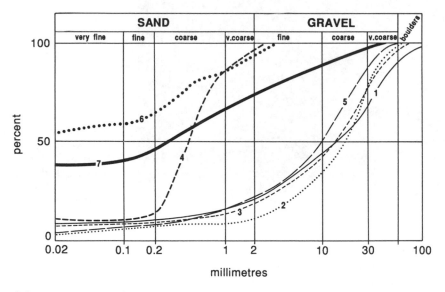

Figure 5.6
Fluvial sediments of the underground River Rak in Planinska Jama, Slovenia. A sediment section in a former phreatic passage has been analysed in detail; from these particle-size data the former water velocities in the passage can be reconstructed. After Gospodaric (1976).

samples are poorly sorted (low gradient of the cumulative curves in figure 5.6), sample 4 is well sorted, and this may reflect either a change in provenance or unusual persistence in river flow. Thus some information may be gained about both magnitude and persistence of river flow. Alternative approaches to this problem involve scanning electron microscopy of included quartz grains (Bull et al. 1989) or estimation of flow velocity from erosional scallops (Curl 1974).

Diagenesis of Cave Sediments

Once emplaced into a cave, any allogenic sediment is existing under conditions of total darkness, near constant high humidity, and near constant temperature. This reduces the amount of chemical alteration that can occur. However, with time some migration of solutes into and out of the sediment will occur. This may be as the result of wetting and drying cycles owing to floods. In this context the porosity and mean grain size of the sediments has a great influence on the degree of diagenesis. Blue and red banded clay sediments in Selminum Tem, New Guinea (see plate 5.4), owe their banding to alternate layers of very fine clay and slightly coarser silt. Reducing conditions are maintained in the fine clays, with a dominance of ferrous iron salts, while in the silts the increased porosity allows oxidation and a dominance of ferric iron. Sediment banks are commonly cemented by iron oxide bands or by calcite from drips. Truly ancient sediments have reaction rims, often of calcite or phosphatic minerals, which may penetrate into the adjoining rock surfaces.

Stratigraphy and its Interpretation

Relationships between cave sediment structures and depositional energy exist, and can be used to reconstruct past discharges. Interactions between cave sedimentation and passage hydrology are relevant. Reconstruction of cave hydrology from sediment surface textures using scanning electron microscopy is an important recent development.

When we come to the task of interpreting the flow conditions under which a sediment was deposited, we can approach the problem in two ways: by using sediment transport theory, which relates the size of individual sediment particles to velocities of erosion, transport and deposition, or by using either empirical or theoretical relationships between sedimentary structures and the hydrologic regime. The former method produces more quantified results but neglects the effects of bed roughness at the sediment–fluid interface. The latter method is not so amenable to quantification but provides more information about the turbulence of flow, the frequency of

Plate 5.4
*An abandoned river
canyon in Selminum Tem,
Papua New Guinea, has
deep mudflows and
waterlain sediment mounds
capped by speleothems
dated by uranium series
methods to 15 000 years
BP.*

deposition and reworking of the sediments. Both approaches have
their adherents, and probably the best solution is to use elements of
each one to achieve the most complete statement possible.

Sediment Transport and Particle Size

The approximate velocity of flow at the moment of deposition can
be estimated from the mean or modal particle size of the sediment.

Waterlain clastic sediments in caves range from boulders to fine clays. The particle-size distribution is often skewed to the fines because of the winnowing or filtering effects of transport though cave conduits. They are generally well sorted and often show a considerable degree of particle flattening (especially the gravels). Thus they do not behave as the spherical particles implicit in sediment transport theory. Nevertheless this theory can be applied and provides sensible results.

The theory of sediment transport is complex (Richards 1982) and relies on fluid dynamics theory. Stream power at the sediment bed is a function of several factors:

$$W = rgQq \qquad (5.1)$$

where W is the gross stream power, r the fluid mass density, g the gravitational acceleration, Q is the velocity in $m s^{-1}$ and q the channel slope (Bagnold 1966). This is related to shear stress by the equation

$$w = W/W = t_o n \qquad (5.2)$$

where w is specific stream power, W is stream width, t is mean shear stress ($kg m^{-2}$) and n is stream velocity ($m s^{-1}$).

Sediment is entrained when the shear stress at the bed exerted by the flow exceeds the gravitational and cohesive forces holding the individual particle in place. For spherical particles the critical shear stress is

$$t_{crit} = 0.06(r_s - r)gD \qquad (5.3)$$

where D = particle diameter in mm and r_s is the particle density.

The maximum size of sediment particle moved in a given channel is given by

$$D_{max} = 65t_o^{0.54} \qquad (5.4)$$

For deposition, the settling velocity n_t is given by Stokes's law:

$$n_t = 1/18 \frac{(r_s - r)g}{m} D^2 \qquad (5.5)$$

where m is the dynamic viscosity. Thus for a given settling velocity the mean size of particles which will fall from suspension can be found. Once entrained, several modes of transport are possible:

- rolling of grains on a stationary bed
- saltation of grains above the bed
- sliding bed with varying depths of particle movement
- suspension load.

Whether sediment is transported by each or all of these modes will depend on mean grain size, on the persistence of flow velocity, and on the roughness of the bed. Each will tend to produce distinctive sedimentary structures which provide an opportunity for later analysis.

An alternative approach is to use the empirical relationships between velocities of flow, transport and sedimentation developed by Hjulström (1935). In figure 5.4 these relationships are displayed according to mean grain size of the particles. The velocity of transport is always less than the velocity of erosion, as more hydraulic force is needed to tear the particles away from the cohesive bed of sediment than is needed to keep them in suspension or saltation. The critical velocities of erosion for fine particles such as silts and clays are as great as those required for coarse sands and gravels. This is owing to cohesive forces between the fine grains, and the smooth surface of the sediment bed which reduces turbulence. Thus it is very hard to erode fine clays, but once eroded they can be kept in transport very easily. In contrast sands are easily eroded and transported, and their velocities of sedimentation are not much less than their critical erosion velocities. Thus sands tend to move through a cave system in a series of hops from temporary storage to temporary storage, and may move right out of the cave altogether. To some extent this explains the markedly bimodal nature of cave sediments, in which gravels and muds predominate.

If we consider the particle size distributions of selected cave sediments, then it is apparent that many lack a single central peak. Cave sediments tend to be skewed, bimodal or polymodal in their particle-size distributions. Using a measure of central tendency such as the mean grain size is perhaps misleading, and it is probably better to employ one or more of the modes. Clayey gravels display bimodal distributions, and act like 'plum puddings' in that the surface roughness caused by protruding gravel particles enhances the erosion of the fine clays by locally increasing flow velocity. Thus there are important factors in sediment erosion and transport which are not accounted for by sediment transport theory.

For this reason we must look at cave sediment structures if we wish to understand their history fully. The best way to do this is to cut a vertical face on a sediment bank and examine the layering carefully. The thickness and inclination of sediment layers will tell much about the duration and energy of flow in a cave passage, while the disruption of those layers by slumping or by erosion will provide information on fluctuations in flow which may correspond to seasonal or longer-term perturbations in the karst system. The most difficult thing to determine is whether a thick bed of sediment is the result of one or many flow events in the cave. The only way to resolve this, apart from a very careful examination of the sediment structures, is by high resolution dating of a number of layers in the bed. This may be possible using the record of secular variation in the earth's magnetic field (Noel and Bull 1982). It is unlikely that radiometric methods will have the required temporal resolution.

There are relationships between water flow and depositional

energy as expressed in clastic sediment structures (see figure 5.7). These can be organized along gradients from low to high energy, and from pools of deeper water to shallower cave streams (unconfined to confined channels). Where deep, virtually static water bodies occur – after floods or in the phreatic zone – sedimentation can result on very steep slopes of up to 70° by the slow settling of colloidal clay particles (see plate 5.5). These sediments accrete parallel to underlying rock or sediment surfaces (Bull 1981) and are often known as cap muds. Because of their mode of deposition, they have the potential to align with the prevailing geomagnetic field, and can thus provide a record of secular variation in magnetic declination and inclination (Noel and Bull 1982). At higher energy levels there is dumping of pulses of sediment into cave pools, producing inclined beds of sediments analogous to prograding deltas. At very high levels – those associated with flash flooding – the extreme turbulence connected with pulsed flow produces curled flame structures in the finer sediments. Rapid slumping of sediments on the deltaic slope may also occur in ways analogous to deep-sea turbidity currents.

Figure 5.7
Scheme of fluvial cave sediment structures in relation to depositional energy and passage morphology. From Gillieson (1986).

Where the flow of water becomes more confined, as in cave streams, then there is a sequence of sedimentary structures ranging from horizontally stratified muds and sands to increasing degrees of

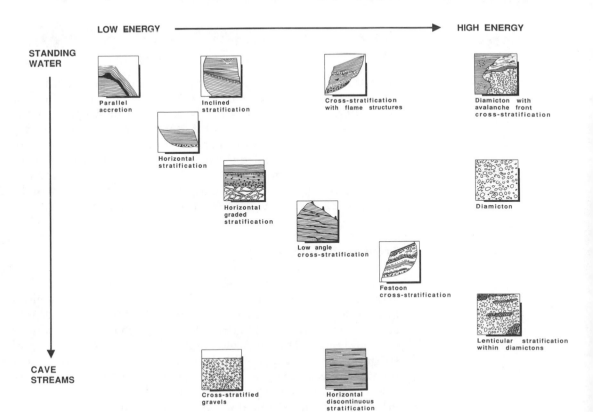

Plate 5.5
Imbricated cave gravels
nearly fill the inverted
siphon of Murray Cave,
Cooleman Plain, Australia.

cross-stratification. With growing energy levels, upward grading of beds from coarse to fine becomes common. At high, turbulent flow conditions, scouring of the bed associated with standing ripples and waves produces lensed cross-stratification. These features are very obvious in cross-section (see figure 5.7) and allow interpretation of the hydraulic regime under which deposition occurred as well as the degree of confinement of water flow. In combination with sediment transport theory, these allow a fairly complete reconstruction of the processes leading to the deposition of a cave sediment.

Provenance Studies

Cave sediments have the potential to provide us with a record of the geomorphic history of their catchments. Analysis of the mineralogy of the gravel and sand fractions of cave sediments permits identification of sediment source areas when the catchment has distinctly varied lithology. Thus, in Wales, Bull (1978) related the lithology of stream gravels in Agen Allwedd to autogenic sources (primarily limestone breakdown) or allogenic sources at the cave extremities. These mixed gravels showed significant downstream changes in lithological composition, and these changes could be related to the geomorphic evolution of the cave system. Magnetic minerals could also be used as tracers in this way.

A very recent approach to this problem uses environmental radionuclides such as ^{137}Cs, ^{226}Ra and ^{232}Th to provide a signature for allogenic sediments entering the Jenolan Cave system, New South Wales (Stanton et al. 1992). Recent sedimentation within the tourist section of this long cave system has led to periodic inundation of walkways and deposition of fines on flowstones and stalagmites. Concentrations of the fallout radionuclide ^{137}Caesium in these recent sediments indicated that most of the deposition had occurred since the mid-1950s. The ratio of ^{226}Radium to ^{232}Thorium in the recent cave sediments was compared to values for soils in different subcatchments upstream of the caves; the results indicated that the recent cave sediments were derived from several catchments in which forestry activities had commenced during the early 1950s. The most likely source was erosion from unsealed forestry access roads in steep terrain.

Thus cave sediment studies have the potential to assist land managers in identifying sources of accelerated erosion in karst terrains, and in reconstructing the long-term hydrology of karst catchments. There is an urgent need for detailed process research into the mechanisms of transport, deposition and diagenesis of cave sediments, especially during flood events. Scanning electron microscopy may offer a very good chance of relating sediment features to depositional energy, but this will require use of recirculating flume experiments as well as within-cave studies.

Caves and Flood History in the Kimberleys, Australia

The estimation of extreme flood discharges is of practical importance for the design of hydrological structures, especially for dams and for flood mitigation works that require information on rare events. A relatively new method uses palaeoflood techniques to estimate the probable maximum flood. In outline, palaeoflood techniques are based upon the occurrence of slackwater deposits (SWD) in bedrock gorges. These are deposited at sites sheltered from high velocity flows, and ideally form sedimentary sequences in which individual flood events can be discerned. Suitable sites occur in limestone caves and at the junctions of narrow side gorges. Once such deposits are located they are described in section and the site surveyed to establish height above the river channel. Suitable material is collected for dating by various absolute techniques, principally radiocarbon and thermoluminescence. The next step is to model the discharges corresponding to the slackwater deposits. The requirements for such flood reconstructions are detailed surveyed cross-sections and field estimates of roughness, as Manning's n values. The calibration of such models in palaeoflood hydrology is

Plate 5.6
The sheer walls of Geikie Gorge in the Kimberleys record a huge flood in 1986 which cleaned the limestone of algae and lichens to a height of 15 m above dry season water level.

Plate 5.7
A deep slackwater deposit of sands and organic silts infills a limestone incut in Windjana Gorge, Kimberleys. This deposit has an associated TL age of 2000 years BP.

WINDJANA GORGE FLOW SIMULATION

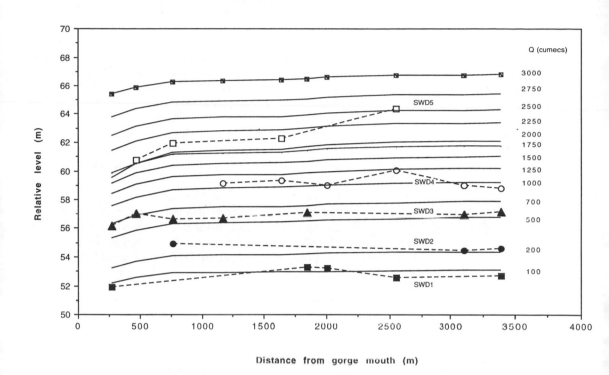

Distance from gorge mouth (m)

aided if detailed information is available for a recent flood debris line.

In the Kimberley region of Australia the hard reefal limestones of the Napier and Oscar ranges have been exhumed from a cover of Permian sandstone. Windjana Gorge has been cut through the Napier Range by the Lennard River, which drains approximately 1200 km² of remote sandstone terrain in the King Leopold Range to the north of the gorge. Numerous caves and narrow side gorges preserve flood deposits. The Kimberleys are located in the monsoonal region of northern Australia. On average 85 per cent of the annual precipitation of some 650 mm falls in the period December to March. The rivers are characterized by very large wet season flows (see plate 5.6), which have dire consequences for bridges, roads and housing. Stream gauging records are available only for the last twenty years and are of variable quality.

Five distinct slackwater deposits were identified in Windjana Gorge (Gillieson et al. 1991). The flow simulation of Windjana Gorge indicates that the lowest two slackwater accumulations (SWD1 and 2) are recent and fall within the range of the mean annual flood. In the last five hundred years only six slackwater accumulations have survived at these two lowest sites. These are

Figure 5.8
Flow simulation and palaeoflood reconstruction for slackwater deposits in Windjana Gorge, Kimberleys karst, Western Australia. From Gillieson et al. (1991).

associated with minimum discharge estimates of 100 to 200 m³ s⁻¹.
SWD4, dated at *c*.2000 years BP (see plate 5.7), is associated with
a discharge estimate of *c*.1000 m³ s⁻¹; this discharge has been exceeded several times in the last decade. SWD5 is dated at *c*.2800
years BP by thermoluminescence and has an associated discharge
estimate of *c*.2600 m³ s⁻¹; this is larger than the maximum recorded
flood of 1986. Thus the stratigraphic record suggests that only one
flood in the last two thousand years has equalled the 1986 flood,
and that only one flood has exceeded it in the last three thousand
years (see figure 5.8).

One approach to the estimation of the magnitude of discharges
from very rare floods is to consider the discharge per km² of
catchment. However, figure 5.9 does show the envelope curve for
the Q_{100} (discharge for the 1 in 100-year event) for Australia and for
'the largest discharges recorded in the world'; they are approximately coincident. The maximum discharge estimated from the
SWD of Windjana Gorge is 2600 m³ s⁻¹. This gives a discharge per
km² of some 2.2 m³ s⁻¹ km⁻². This is a high discharge per unit area
but falls below both the Australian Q100 and world maximum

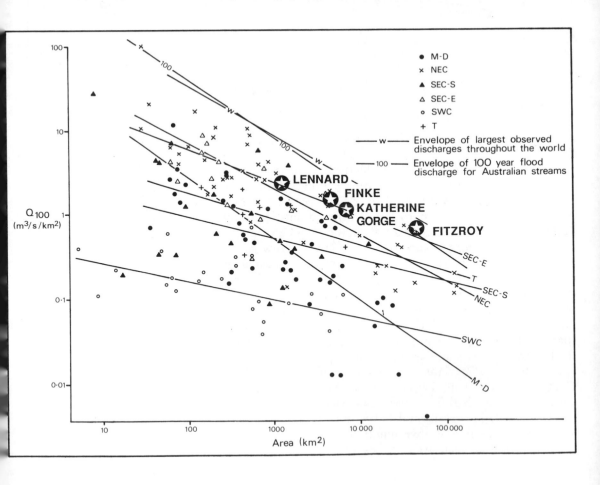

data. Other northern Australian rivers analysed in this way are the
Katherine Gorge in the Northern Territory ($Q100 = 0.59\,m^3\,s^{-1}$
km^{-2}) and the Fitzroy River of Western Australia ($Q100 = 0.65\,m^3$
$s^{-1}km^{-2}$). Thus the Lennard River palaeoflood reconstruction yields
an areal discharge that is high by regional standards.

There is a growing interest in the contribution to be made by
palaeoflood studies, as they can provide an additional and inde-
pendent technique for the estimation of rare and extreme flow
events. Cave sediments in limestone gorges can offer a unique
source of data for these reconstructions.

References

Allen, J. R. L. 1985: *Physical Processes of Sedimentation*, 2nd edn. London: Allen & Unwin.

Bagnold, R. A. 1966: *An Approach to the Sediment Transport Problem
from General Physics*. US Geol. Survey Prof. Paper 422–1.

Bögli, A. 1980: *Karst Hydrology and Physical Speleology*. Berlin: Springer.

Bull, P. 1978: A study of stream gravels from a cave: Agen Allwedd, South
Wales. *Zeitschrift für Geomorphologie* 22 (3), 275–96.

Bull, P. A. 1981: Some fine-grained sedimentation phenomena in caves.
Earth Surface Processes and Landforms 6, 11–22.

Bull, P. A., Yuan, D. and Hu, M. 1989: Cave sediments from Chuan Shan
tower karst, Guilin, China. *Cave Science* 16, 51–6.

Collier, C. R. and Flint, R. F. 1964: Fluvial sedimentation in Mammoth
Cave, Kentucky. *US Geological Survey Professional Paper* 475D, D141–
D143.

Curl, R. L. 1974: Deducing flow velocity in cave conduits from scallops.
Bull. Nat. Speleo. Soc. Am. 36 (2), 1–5.

Frank, R. M. and Jennings, J. N. 1978: Development of a subterranean
meander cutoff: the Abercrombie Caves, New South Wales. *Helitite* 16,
71–85.

Gascoyne, M., Ford, D. C. and Schwarcz, H. P. 1983: Rates of cave and
landform development in the Yorkshire Dales from speleothem age data.
Earth Surface Processes and Landforms 8, 557–68.

Gillieson, D. 1985: Geomorphic development of limestone caves in the
Highlands of Papua New Guinea. *Zeitschrift für Geomorphologie* 29,
51–70.

Gillieson, D. 1986: Cave sedimentation in the New Guinea highlands.
Earth Surface Processes and Landforms 11, 533–43.

Gillieson, D., Smith, D. I., Greenaway, M. and Ellaway, M. 1991: Flood
history of the limestone ranges, Kimberleys, Western Australia. *Applied
Geography* 11, 105–24.

Gospodaric, R. 1976: The Quaternary caves development between the
Pivka basin and Polje of Planina. *Acta Carsologica* 7, 5–135.

Hjulström, F. 1935: Studies of the morphological activities of rivers as
illustrated by the River Fyris. *Bull. Geol. Inst. Univ. Uppsala* 25, 221–
527.

Ikeya, M., Toshikatsu, M. and Gospodaric, R. 1983: ESR dating of
Postojna Cave stalactite. *Acta Carsologica* 11, 117–30.

Kranjc, A. 1981: Sediments from Babja Jama near Most na Soci. *Acta
Carsologica* 10 (9), 201–11.

Noel, M. and Bull P. A. 1982: The palaeomagnetism of sediments from Clearwater Cave, Mulu, Sarawak. *Cave Science* 9 (2), 134–41.

Richards, K. 1982: *Rivers: Form and Process in Alluvial Channels.* London: Methuen.

Schroeder, J. and Ford, D. C. 1983: Clastic sediments in Castleguard Cave, Columbia Icefields, Alberta, Canada. *Arctic and Alpine Research* 15 (4), 451–61.

Siffre, A. and Siffre, M. 1961: Le façonnement des alluvions karstique. *Ann. Spéléo.* 16, 73–80.

Smart, P., Waltham, T., Yang, M. and Zhang, Y. 1986: Karst geomorphology of western Guizhou, China. *Trans. Brit. Cave Res. Assoc.* 13 (3), 89–103.

Stanton, R. K., Murray, A. S. and Olley, J. M. 1992: Tracing recent sediment using environmental radionuclides and mineral magnetics in the karst of Jenolan Caves, Australia. In J. Bogen, D. E. Walling and T. J. Day (ed.) *Erosion and Sediment Transport Monitoring Programmes in River Basins.* Proceedings of a symposium held at Oslo, August 1992. IAHS Publ. no. 210. Wallingford, UK: IAHS Press.

Valen, V. and Lauritzen, S. E. 1990: The sedimentology of Sirijorda Cave, Norland, northern Norway. *Proc. 10th Int. Congr. Speleo., Budapest,* 125–6.

Williams, P. W., Lyons, R. G., Wang, X., Fang, L. and Bao, H. 1986: Interpretation of the palaeomagnetism of cave sediments from a karst tower at Guilin. *Carsologica Sinica* 6, 119–25.

SIX

Dating Cave Deposits

The Importance of Dating Cave Deposits

Most of the conceptual advances in the scientific study of caves have derived from the application of new dating techniques to cave deposits such as calcite and gypsum speleothems, and to layered sediments. This development has occurred at a rapid pace over the last two decades, and new analytical methods are steadily appearing as a result of cooperation between geomorphologists, geologists and physicists. Whereas twenty years ago cave development was largely constrained into the Last Glacial–Interglacial cycle, at the end of the 1990s the long history – well into the Tertiary in some cases – of caves is being recognized. This has far-reaching consequences for our perception of rates of geomorphic process operation, and for the construction of schemes of regional denudation chronology.

The dating of cave deposits has wider import than merely providing a chronology for the clastic sediments and their included flora and fauna, long the preserve of archaeologists and palaeontologists. While cave sediment chronology remains an important area of speleology, there is keen interest in finding the minimum ages of cave passages. Usually the dating of speleothems or sediments will only give us an age before which a passage must have been formed. In some cases, new passages will have formed after a deposit was laid down; this provides a maximum age for the passage. If the oldest deposits in a vertical sequence of caves also form a time sequence (see plate 6.1), then the time of formation of individual passages can be more precisely determined. If those passages are graded to a present or former river valley or erosion surface, then we can infer rates of landscape evolution from the sequence of cave passages and their deposits. This is perhaps the most powerful application of cave chronology, as here the caves are preserving evidence of past landscape development which has elsewhere been obliterated by surface processes. Such applications have radically revised thinking about the age and rate of evolution of landscapes

Plate 6.1
*Daxiao Shan, a
perched phreatic
tunnel north of
Longgong, Guizhou,
China, is testimony to
rapid uplift and valley
deepening on the
margin of the
Himalayas. The
passage is 50m high
and wide and 800m
long.*

as diverse as the Yorkshire Dales of England and the tower karst of
lowland tropical China.

Dating Techniques and the Quaternary Timescale

At an early stage in any speleological investigation, decisions must
be made about the dating techniques to be employed. Those decisions will rest on the perception of the likely age of the cave
passages and their deposits, on the availability of funds or access to
a dating laboratory, and on the questions being asked. Thus a
geologist interested in the long-term evolution of a landscape will
likely use a method with a potential time range of millions of years,
while an archaeologist concerned with patterns of human occupation of a site since the Last Glacial will probably opt for the
radiocarbon technique. The time range of several widely used techniques is given in figure 6.1. One obvious feature is the big gap in
available techniques in the time range from $c.400\,000$ years to 1
million years. Unfortunately we are today finding many sites whose
antiquity falls in or beyond that gap. The challenge for the future is
to devise, test and apply new dating methods that can cope with the
time range from 10^4 to 10^6 years. We can divide the currently
available techniques into three groups:

- *comparative,* where a preserved physical or chemical
  parameter of the sample is compared with values of the
  same parameter in a dated sequence – for example,
  palaeomagnetism

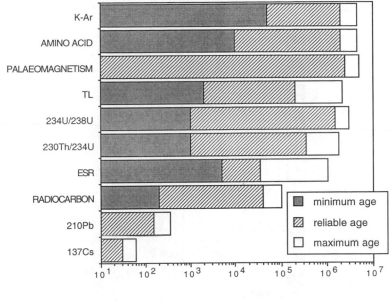

Figure 6.1
The Quaternary timescale and the effective range of comparative, radiometric and radiogenic dating methods.

- *radiometric*, where the statistically random but predictable decay of a radioisotope incorporated into the sample is measured – for example, uranium-series and radiocarbon methods
- *radiogenic*, where the crystal defects created in a sample by the radioactive decay of an isotope are measured – for example, electron spin resonance (ESR) and thermoluminescence (TL) dating.

Palaeomagnetism

One technique which does cover the time range from tens of thousands to a few million years is the comparative method of palaeomagnetic analysis. Unfortunately it is a low precision method, in which individual samples can only be placed into broad time classes in which the magnetic polarity of the material is either normal or reversed. This magnetic timescale (see figure 6.2) extends back in time for the last three million years, i.e., the whole Quaternary, although deep-sea drilling extends the method almost yearly.

So far most success has been achieved by analysing the remanent magnetic field preserved in fine-grained clastic sediments (pond deposits) within caves. These settle slowly in accordance with the prevailing geomagnetic field, although the dip of individual particles is some 30° shallower than the earth's field on account of final random motions on irregular sediment banks. In the cave environ-

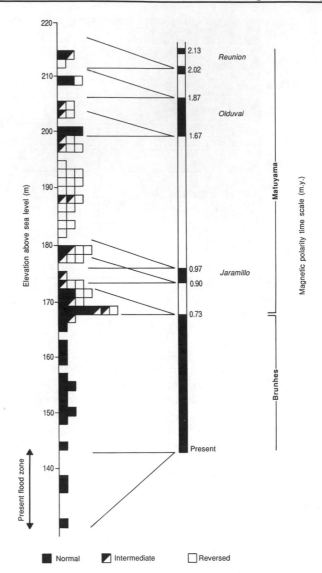

Figure 6.2
The magnetic polarity timescale and the record of normal and reversed magnetic declinations obtained from samples of sands, silts and clays throughout the Mammoth–Flint Ridge–Roppel–Proctor Cave system of Kentucky. From Schmidt (1982).

ment these pond deposits are little affected by chemical change, although some oxidation may occur and calcite deposition may take place under drips. These secondary effects have their own magnetic field, which can usually be 'cleaned' from the specimen during analysis in the laboratory.

In most cases the results are compared with the geomagnetic record of reversals, such as in the Mammoth–Flint Ridge–Roppel–Proctor Cave system of Kentucky (Schmidt 1982). In figure 6.2 the record of normal and reversed magnetic declinations of fine-grained sediments from this vast cave system are related to the magnetic polarity timescale. The lowest and youngest passages, related to the progressive downcutting of the Green River, have normal polarity. Samples above 180 m elevation in the cave (above Gothic and Grand avenues) have reversed polarity and probably predate the Brunhes–Matuyama transition at 730 000 years. There is a suggestion that the highest passages (above 200 m elevation, the level of Collins Avenue) may date to the normal polarity Olduvai epoch at 1.6 million years, or slightly older. A problem with this approach is the patchy nature of the cave magnetic record; it is punctuated by large gaps, unlike deep-sea cores. Thus a reversal may be a result of deposition in any of several epochs. Only where a large number of samples has been taken – in the above case, five hundred oriented samples of clays, sands and silts – can any confidence be attached to age estimates.

Magnetostratigraphy can also be used to make inferences about regional denudation chronology, as cave systems may encapsulate elements of the geomorphic history of the overlying landscape. Pease et al. (1994) have studied the polarity of sediment samples from the multi-level Wyandotte Cave, southern Indiana. This system lies to the north of Mammoth Cave and, like that cave system, its base level has also been a tributary of the Ohio River. The sediments in the uppermost levels of Wyandotte Cave have a normal polarity, while those lower in the sequence have reversed polarity and correlate with old tube passages such as Upper Salts Avenue, Grand Avenue and Cleaveland Avenue. These passages were thus active not less than 780 000 years BP (Brunhes–Matuyama transition). The uppermost passages are infilled by sediments not less than 2.6 Ma old (Matuyama–Gauss transition), and these correlate with the residual sediments of the Upper Mitchell Plain of Indiana and the Pennyroyal Plateau of Kentucky. This provides a minimal age for this widespread land surface that has controlled karst development.

In truly ancient materials the preserved magnetic field may be compared to the known position of the geomagnetic pole (the Apparent Polar Wandering Path) over geological time, and an age estimate obtained. If a thick sequence of sediments is present, then samples taken at close depth intervals may be used to construct a record of magnetic secular variation. This can be matched with long records of secular variation gained from analysis of lake sediments or deep-sea sediment cores. In Peak Cavern, Derbyshire, Noel (1986) has matched a one-metre sequence of laminated clays with lake magnetostratigraphy, implying an early Holocene age. Palaeocurrent directions in the cave passage were also obtained. In another study (Noel and Bull 1982), post-depositional disturbance has affected the secular magnetic record of a laminated silt section from Clearwater Cave, Mulu, Sarawak. This may be owing to burrowing by cave crickets.

Calcite speleothems acquire some chemical remanent magnetization during the crystallization process and subsequent alterations. Detrital grains from floodwater may also be incorporated into the speleothem, and may provide a stronger magnetic signal. In either case the strength is low, requiring use of very sensitive magnetometers. Palaeomagnetic dating can thus be used to supplement absolute dating methods: if the sample is older than 350 000 years, then normal or reversed polarity will indicate if it is younger or older than the Brunhes–Matuyama reversal at 730 000 years BP – or another geomagnetic reversal. Although results (including secular variation) have been obtained for speleothems from North America (Latham et al. 1979; Latham et al. 1986), problems with the identity of the speleothem magnetization remain unsolved (Latham and Schwartz 1990).

Radiocarbon

This is an absolute radiometric method relying on the steady decay of naturally occurring radiocarbon or carbon-14. This radioisotope is produced by cosmic ray bombardment of nitrogen molecules in the upper atmosphere. Roughly 1 in 10 000 atoms of carbon in the atmosphere is radiocarbon. Radioactive isotopes are unstable and decay spontaneously, emitting radioactivity. This decay occurs at a set rate for each radioisotope and is unaffected by external influences. The rate is defined by its half-life ($T^{1/2}$), the time taken for half of the atoms to decay following the exponential law

$$N = N_o e^{-\lambda t} \qquad (6.1)$$

where N is the number of atoms present at time t, N_o was the number present at the start of decay, and λ is the half-life or decay constant. For radiocarbon (^{14}C) the half-life is 5730 years; this gives an effective upper limit of six or seven half-lives, or around 40 000 years.

When calcite is precipitated in the cave we assume that the system is closed, that is carbon locked into the crystal lattice cannot subsequently exchange with any later HCO_3 ions passing over the calcite. The decay of radiocarbon thus measures the time since the calcite was precipitated. The carbon in calcium carbonate may come from two sources. Half should come from the bedrock calcium carbonate, where ^{14}C is absent due to its great age. The other 'live' half should come from the air or from humic compounds in the soil organic matter. In reality the proportions are different. For Castleguard Cave in Canada, Gascoyne and Nelson (1983) estimate the proportion of live carbon as 35 per cent, while in Belgian caves it is estimated at 80 to 85 per cent (Bastin and Gewellt 1986). The only method for resolving this problem at a cave site is to measure the isotopic composition of cave drip waters over several seasons, calculate a correction factor, and assume that this also applied in the past. Recently Dulinski and Rosanski (1990) have developed a model which adequately predicts the carbon isotope composition of cave deposits from a consideration of chemical kinetics, but this will require extensive testing before its widespread adoption. For this reason, and the short time range afforded by the technique, radiocarbon age determinations on speleothems have been very limited in scope (Hendy 1970; Heine and Geyh 1984). The method has, however, been extensively used to date organic matter in cave entrance deposits (see plate 6.2), primarily archaeological sites (Gillieson and Mountain 1983). More recently radiocarbon has been successfully used to provide a late Pleistocene and Holocene chronology for stalagmites in Drotsky's Cave, Botswana

Plate 6.2
The ancient river cave at Zhoukoudian, China, was infilled by fluvial and hillslope sediments, including human remains, up to 1.5 million years old. This fill has been excavated by several generations of archaeologists.

(Burney et al. 1994). A good correlation between uranium series and radiocarbon ages was obtained from a sixty-centimetre core taken from two coalesced formations, and extraction of the pollen preserved in the growth rings has provided a post-glacial vegetation history for this site on the edge of the Kalahari Desert. However, Holmgren et al. (1994) have also carried out a similar analysis on stalagmite from Lobatse Cave in Botswana; they note a good correlation between uranium series and radiocarbon ages younger than 20000 years, but before that the relationship breaks down, with a discrepancy of 20000 years for older material with a uranium series age of 50000 years.

Uranium series

The radiometric method known as uranium series disequilibrium is the most widely used and most successful for the dating of calcite speleothems. It has an effective range from a few thousand to 350000 years BP. The first successful uranium series dates on speleothems were obtained by Duplessy et al. (1970) on material from the Aven D'Orgnac, France. Since then several other laboratories have been established, notably in Canada, Great Britain, Russia, Japan, New Zealand and Australia; by far the greatest number of age determinations (well over a thousand) have been produced by the McMaster (Canada) laboratory under the leadership of D. C. Ford and H. P. Schwarcz (Gascoyne 1984).

The primary source of uranium isotopes is the weathering of igneous rocks and the incorporation of the weathering products in the hydrological cycle. Two natural parent isotopes, ^{238}U and ^{235}U, exist. These have exceptionally long half-lives, and decay by emission of α and β particles to stable lead isotopes ^{206}Pb and ^{207}Pb. Intermediate daughter products such as ^{234}U and radon (^{226}Ra) are also suitable for dating. Upon weathering, more daughter ^{234}U is released than the parent ^{238}U and ^{235}U. All three species become oxidized and are readily transported in solution as complex ions (see figure 6.3). They may be precipitated in calcite. The daughter products proactinium (^{231}Pa) and thorium (^{230}Th) are insoluble and may bond to the charged surface of clay particles or organic free radicals. They will be present in the calcite only as detrital layers or impurities. Thus, in a pure calcite, any ^{230}Th will be the result of decay of ^{234}U over time since crystallization. This forms the basis of the uranium series method as applied to speleothems. A complex equation (Gascoyne et al. 1978) governs the decay of ^{234}U to ^{230}Th. This is summarized in figure 6.4. The initial ratio of ^{234}U to ^{238}U is normally always greater than 1.0 and the ratio of ^{230}Th to ^{234}U is greater than 0. Through time ratios evolve to the right, to a point where the steepening isochrons cannot be resolved. This is at an age around 350000 years. For samples in excess of 350ka, where the initial ratio is unknown, an estimate of the age can be obtained by assuming an initial $^{234}U/^{238}U$ ratio equal to the regional mean. This

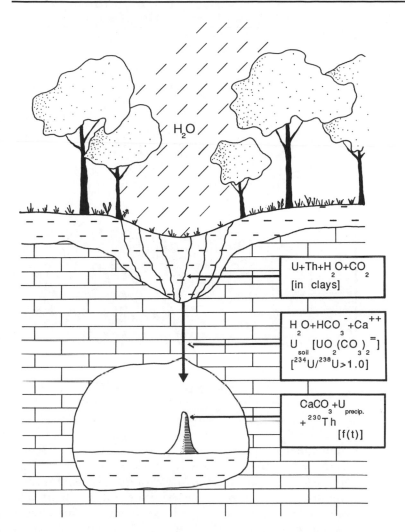

Figure 6.3
Schema of the geochemical pathways of uranium and thorium into caves and speleothems.

is the Regional Uranium Best Estimate (RUBE) method of Gascoyne et al. (1983), which permits age estimates up to 1.5 million years.

Several criteria must be met for reliable uranium series dating of speleothems:

1   There must be sufficient uranium in the calcite. The absolute minimum is 0.01 ppm, while 90 per cent of samples have more than this concentration, up to a recorded maximum of 120 ppm. At some sites where speleothem growth is dependent on direct rainwater input only, uranium concentrations may be too low.

2   The calcite must be closed to exchange after the coprecipitation of calcite and uranium salts. This assumption is frequently violated, and careful examination of

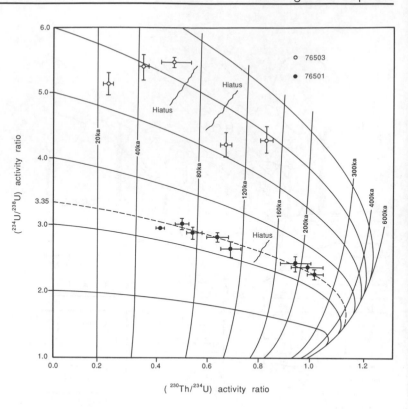

Figure 6.4
Graphical illustration of the $^{230}Th:^{234}U$ dating method. Two speleothems (76501 and 76503) have activity ratios of 3.35 and 5.0–6.0 respectively; the radiometric evolution of the first shows a near-constant activity ratio through time, while the second shows considerable variation with a hiatus. This latter effect may be owing to exchange with the surrounding environment. After Ford and Williams (1989).

individual samples for areas of porous calcite, of resolution or of reprecipitation is necessary. For this reason porous tufas and stalactites are avoided. This entails careful sampling and sectioning, with the real possibility that a sample from a geomorphologically meaningful context will have to be rejected. There is also a question of conservation involved, as any sampling is destructive to the cave environment. Time spent in personal reconnaissance is seldom wasted.

3   No ^{230}Th or ^{231}Pa must be deposited in the calcite. In reality most calcites contain some of these insoluble isotopes as contaminants. Usually the concentration of ^{232}Th is measured in the sample, an initial ratio of $^{230}Th/^{232}Th$ assumed, and a correction applied for allogenic ^{230}Th. New, more sophisticated methods use partial leaching and assay techniques (Latham and Schwartz 1990). If the ratio of $^{230}Th/^{232}Th$ is greater than 20, radiogenic thorium is assumed to dominate and contamination is minimal. Highly contaminated specimens are either avoided or subjected to repeat determinations. The latter is a costly process. The major sources of detrital ^{230}Th are clays,

deposited from floodwaters, organic colloids in soil, or the fine carbon particles of bushfire smoke. At Yarrangobilly in the Snowy Mountains of Australia, thin black layers in flowstones are made of carbon dust which contains abundant thorium liberated from the soil during the intense bushfires which burn the area once a decade or so.

Recent developments in uranium series dating using isotope dilution mass spectrometry (Li et al. 1989) offer the prospect of reliably dating speleothems with U contents as low as 0.08 ppm to 400 000 years. The new method offers high precision with counting errors reduced significantly. For material with higher U content (>3 ppm) it should be possible to date material as old as 600 000 years. The problem of detrital thorium remains, however.

The application of uranium series dating to the problem of landscape denudation is well illustrated by data from the Yorkshire Dales. Here Gascoyne, Ford and Schwarcz (1983) have analysed a large number of speleothems from nine major cave systems in the Craven district (see figure 6.5). The upper levels of the Ease Gill,

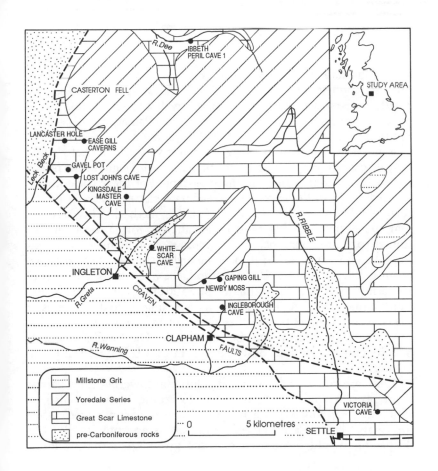

Figure 6.5
The Craven district of northwest England, with locations of caves dated by uranium-series methods. From Gascoyne et al. (1983).

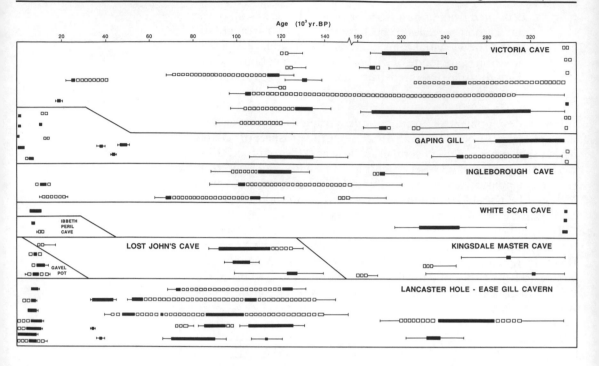

Figure 6.6
Bar graph of 230Th/234U ages and growth periods of speleothems from Craven caves. Error bars are ±1 of ages nearest to base and top of speleothems. From Gascoyne et al. (1983).

West Kingsdale and Gaping Ghyll systems, characterized by relict phreatic tunnels, contain speleothems whose age is in excess of 350 000 years (see figure 6.6). In the well-known Gaping Ghyll system, samples from Mud Hall and Far East Passage are older than 350 000 years, while an age of 120 000 years on the relict phreatic passage of the Far Country suggests that it was drained before that time. Most of the visible deposits are recent, with eight of the fifteen speleothems being formed in the Holocene. For those speleothems older than 350 000 years, an age estimate can be made using the Regional Uranium Best Estimate (RUBE) method. This assumes that the initial $^{234}U/^{238}U$ ratio of an ancient formation might be similar to the regional mean ratio (measured on many speleothems over the entire age range) if the standard deviation of the regional mean is low. Using the RUBE method, the ancient speleothems of Victoria Cave have age estimates of >500 000 years, with the extreme values at 1.5 to 1.9 million years. Using these dates, rates of passage entrenchment and valley downcutting (allowing breaching of phreatic passages) can be estimated. The results in table 6.1 suggest maximum downcutting rates of between 20 and 50 mm ka⁻¹.

These are consistent with areal solutional lowering rates gained from process studies. Mean valley deepening rates can also be estimated; for Kingsdale, a rate of 50 mm ka⁻¹ is indicated, while for Chapel-le-Dale a rate of <200 mm ka⁻¹ is appropriate. But this deepening would have been episodic given Quaternary climate

Table 6.1 Speleothem age and elevation data with calculated passage entrenchment rates for caves in the Craven district, northwest England (from Gascoyne et al. 1983)

Cave	Deposit type	Height above stream (m)	Basal age (ka)	Mean maximum downcutting rate (mm ka^{-1})
Lost John's Cave	Wall flowstone	2.5	115	2.2
Kingsdale Master Cave	Resolutioned roof flowstone	11	300	3.7
Ingleborough Cave	Flowstone	4	≥120	≤3.2
Ease Gill Caverns	Loose flowstone	20	240	8.3
White Scar Cave	Loose flowstone	20	≥350	≤5.7

fluctuations. It is clear that the role of the glaciations which affected the area has been overestimated, with lowering between 6 and <24 m per glacial cycle, in contrast to earlier estimates of 20 to 50 m (Brook 1974).

Other dated speleothems tend to cluster in two distinct groups, associated with the Last Interglacial (oxygen isotope stage 5), the Penultimate Interglacial (oxygen isotope stage 7) and the Holocene (see section 'Timing the Ice Ages' below).

Trapped electron methods: ESR, TL and OSL

There are several closely related radiogenic techniques for dating cave deposits which rely on the presence of radiation-induced defects in the crystal lattice of the sample. These defects take the form of electron traps, which hold electrons liberated by the natural α, β or γ radiation of radioisotopes present in the sample or in its surrounding environment. The number of traps will increase to a state of saturation, at which time the rate of decay of traps equals the rate of their formation. There is a constant increase in the number of these defects with time, and for any sample there will be a Total Dose acquired, which is measured in rads. If the rate of defect formation is known (the Annual Dose) then the age of the sample may be simply calculated from the equation:

$$\text{Age (years)} = \frac{\text{Total Dose (Grays)}}{\text{Annual Dose (Grays year}^{-1})} \qquad (6.2)$$

In most studies the accumulated total dose is determined by the additive technique. In this, duplicate samples are given increasing doses of γ radiation, the response is measured, and the results are subjected to regression analysis through zero (see figure 6.7). From this the pre-irradiation total dose can be determined. For the ESR

(Electron Spin Resonance) method, the number of defects (total dose) is measured from the paramagnetic signal (the ESR spectrum; see figure 6.8) of the sample. For the Thermoluminescence (TL) method, the electron traps are emptied by heating and the emitted light energy measured as a glow curve (figure 6.8). For OSL (Opti-

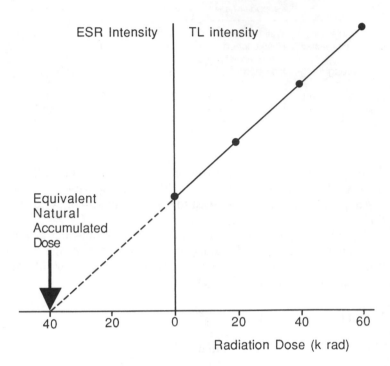

Figure 6.7
The additive dose method of determining the total accumulated dose of radiation in a calcite crystal by means of γ irradiation and regression analysis. Modified from Lyons et al. (1988).

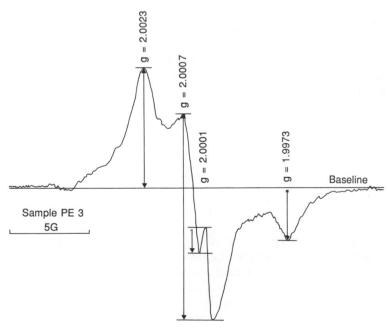

Figure 6.8
a) An ESR spectrum of speleothem calcite from New Zealand. From Lyons et al. (1988);

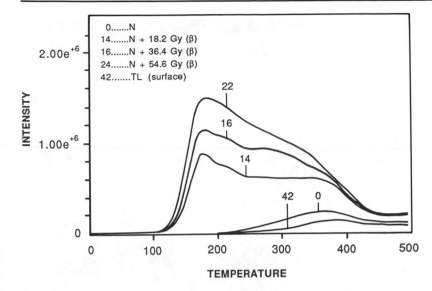

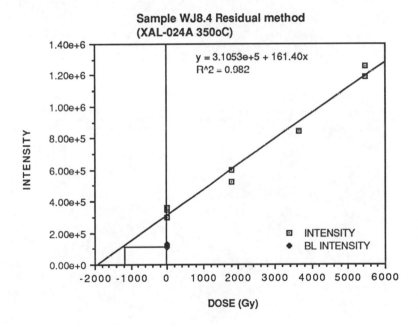

Figure 6.8
b) TL glow curve of quartz sand from cave deposit in Windjana Gorge, Kimberleys, Western Australia. From Gillieson et al. (1991).

cally Stimulated Luminescence), the traps are emptied using a laser beam, and the emitted light again measured. This new technique (Aitken 1990) has the advantage that a narrow energy window is used, individual grains may be analysed, and very fine-grained sediments may be used.

In all methods, the Annual Dose is estimated from the concentrations of radionuclides (U, Th, K) in the sample, or by γ-ray measurement at the site. Determination of this Annual Dose is the most difficult part of the method, and it is not yet resolved

Plate 6.3
The lofty passage of Lower Moon Cave, Guilin, China, is perched 30 m above the Li River which formed it. This passage has been dated by palaeomagnetic analysis of fluvial sediments to at least 730 000 BP.

satisfactorily. In addition, the most appropriate regression equation may not be linear, because of saturation of the electron traps. Correlation between ESR/TL dates and other techniques has been attempted; in some cases the results are excellent, in others very poor. Thus some researchers regard these as relative rather than absolute dating methods, and for now the results must be treated with caution. The ESR method holds great promise for dating speleothems, sands, shells and bone up to 1 million years old; until now most results have come from speleothems (Hennig and Grun 1983; Smith et al. 1985; Lyons et al. 1988). The TL method has been widely applied to quartz sands (Aitken 1985; Quinif 1982) and ceramics, and appears to have an effective range of about 200 000 years.

An example of the combination of dating methods can be seen in the study of the age of tower karst at Guilin, China, carried out by Williams, Lyons and their co-workers (Williams et al. 1986; Williams 1987). Earlier work has suggested that the highest levels of caves (see plate 6.3) in the striking *fenglin* or 'peak forest' karst might be mid-Pleistocene to Pliocene in age, on the basis of fossil fauna. Cave sediments 23 m or more above the floodplain of the Li Jiang river were deposited in a magnetically normal epoch prior to the Brunhes, possibly in the Jaramillo event (900 ka). Some deposits above this level are magnetically reversed. Low-level caves have undergone several phases of burial during floodplain aggradation, followed by partial stripping during incision (see figure 6.9). Thus former floodplain levels are indicated at several elevations. Stalagmites sampled from different cave levels in the same towers as studied by Williams are all beyond the limit of uranium series dating (D. C. Ford, pers. comm.). Bull et al. (1989) have examined sediments from these caves, and conclude that they differ significantly from contemporary river deposits, and that multiple sources are indicated. Sediments may be of fluvial, colluvial (including mudflow) and aeolian origin. Some sediments may even predate the

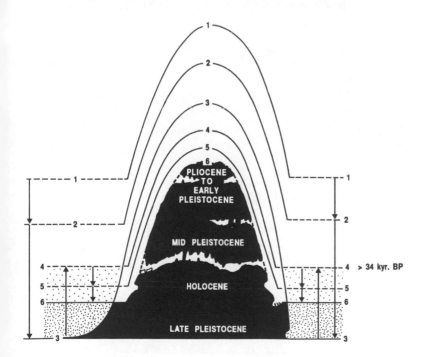

Figure 6.9
Model of karst tower evolution near Guilin, China, with additional data from Williams et al. (1986).

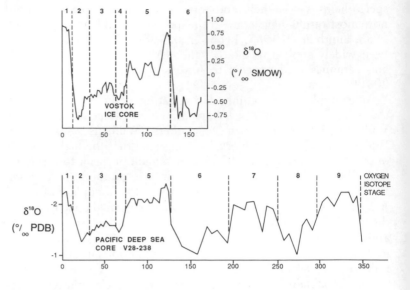

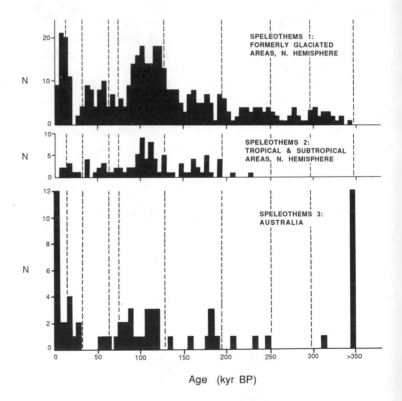

Figure 6.10
Frequency distribution of uranium-series age data for speleothems from North America, Western Europe and Australia. Data from Gascoyne et al. (1983), Hennig et al. (1983), and Goede and Harmon (1983), plus unpublished data of the author.

tower formation. Thus the origin of both the cavities and their infills in this typical tower karst is clearly complex, and their formation has proceeded episodically over a very long timescale.

Timing the Ice Ages

Early attempts to date the Glacials relied on radiocarbon chronologies applied to organic deposits overlying glacial sediments. These provided minimum ages only, and the elucidation of the full glacial sequence was hampered by the fragmentary nature of the evidence. Subsequently the application of stable oxygen isotope analysis to deep-sea cores provided an indirect measure of the timing and intensity of Glacial–Interglacial cycles; long ice cores from the Antarctic have also aided this process of reconstruction. Once a large number of speleothem dates had been assembled for several regions – Canada, Northern England and Tasmania – it was realized that the frequency distribution of speleothem ages reflected the fluctuations in climate (see figure 6.10).

This arose because, during intense glaciation at high latitudes, the karst groundwater would normally be frozen. No stalactites or stalagmites would form until conditions warmed sufficiently to allow groundwater circulation. In the histogram, samples from three widely spaced regions show the same pattern – little or no speleothem deposition during the Glacials and major phases clustered in the Interglacials. Exceptions to this pattern reflect the role of hydrothermal activity or the retarding role of stagnant ice masses left behind by the retreating ice sheets.

Interglacial or interstadial events seem to be before 170 ka, from 140 to 90 ka, at 60 ka and from 16 ka to the present. The Last Glacial Maximum is clearly indicated at 20 ka, and other cold periods are indicated at approximately 160 ka and >250 ka. These older indications are hampered by the low number of dates and the larger error terms associated with them. This may be reduced in future by the use of mass spectrometry. This pattern complements the deep-sea core and ice core records, but is based on better dating control than the former. It provides a reliable chronology for the last 300 000 years, to which other types of records could be matched.

References

Aitken, M. J. 1985: *Thermoluminescence Dating*. London: Academic Press.

Aitken, M. J. 1990: *Science-Based Dating in Archaeology*. London: Longmans, 274 pp.

Bastin, B. and Gewellt, M. 1986: Analyse pollinique et datation 14C de concrétions stalagmitiques Holocènes. *Geog. Phys. et Quat.* 40, 185–96.

Brook, D. 1974: Cave development in Kingsdale. In A. C. Waltham (ed.) *Limestones and Caves of North-West England.* Newton Abbot: David & Charles, pp. 310–34.

Bull, P. A., Yuan, D. and Hu, M. 1989: Cave sediments from Chuan Shan tower karst, Guilin, China. *Cave Science* 16, 51–6.

Burney, D. A., Brook, G. A. and Cowart, J. B. 1994: A Holocene pollen record for the Kalahari Desert of Botswana from a U-series dated speleothem. *The Holocene* 493, 225–32.

Dulinski, M. and Rosanski, K. 1990: Formation of $^{13}C/^{12}C$ isotope ratios in speleothems: a semi-dynamic model. *Radiocarbon* 32 (1), 7–16.

Duplessy, J. C., Labeyrie, C. L. and Nguyen, H. V. 1970: Continental climatic variations between 130000 and 90000 years BP. *Nature* 226, 631–2.

Gascoyne, M. 1984: Twenty years of uranium-series dating of cave calcites: a review of results, problems and new directions. *Studies in Speleology* 5, 15–30.

Gascoyne, M., Ford, D. C. and Schwarcz, H. P. 1983: Rates of cave and landform development in the Yorkshire Dales from speleothem age data. *Earth Surface Processes and Landforms* 8, 557–68.

Gascoyne, M. and Nelson, D. E. 1983: Growth mechanisms of recent speleothems from Castleguard Cave, Columbia Icefields, Alberta, Canada, inferred from a comparison of Uranium-series and Carbon-14 age data. *Arctic and Alpine Research* 15 (4), 537–42.

Gascoyne, M., Schwarcz, H. P. and Ford, D. C. 1978: Uranium series dating and stable isotope studies of speleothems, part I: Theory and techniques. *Trans. Brit. Cave Res. Assoc.* 5 (2), 91–112.

Gillieson, D. and Mountain, M. J. 1983: Environmental history of Nombe rockshelter, Papua New Guinea Highlands. *Archaeology in Oceania* 18, 53–62.

Gillieson, D. Smith, D. I., Greenaway, M. and Ellaway, M. 1991: Flood history of the limestone ranges, Kimberleys, Western Australia. *Applied Geography* 11, 105–24.

Goede, A. and Harmon, R. S. 1983: Radiometric dating of Tasmanian speleothems: evidence of cave evolution and climatic change. *Journal of the Geological Society of Australia* 30, 89–100.

Heine, K. and Geyh, M. A. 1984: Radiocarbon dating of speleothems from the Rossing Cave, Namib Desert, and palaeoclimatic implications. In J. C. Vogel (ed.) *Late Cainozoic Palaeoclimates of the Southern Hemisphere.* Rotterdam: A. A. Balkema.

Hendy, C. H. 1970: The use of C^{14} in the study of cave processes. In I. U. Olson (ed.) *Radiocarbon Variations and Absolute Chronology.* New York: Wiley, 419–43.

Hennig, G. J. and Grun, R. 1983: ESR dating in Quaternary geology. *Quaternary Science Review* 2, 157–238.

Hennig, G. J., Grun, R. and Brunnacker, K. 1983: Speleothems, travertines and paleoclimates. *Quaternary Research* 20, 1–29.

Holmgren, K., Lauritzen, S. E. and Possnert, G. 1994: $^{230}Th/^{234}U$ and ^{14}C dating of a Late Pleistocene stalagmite in Lobatse II cave, Botswana. *Quaternary Geochronology* 13, 111–19.

Latham, A. G. and Schwarcz, H. D. 1990: Magnetization of speleothems: detrital or chemical? *Proc. 10th Int. Congr. Speleo. Budapest.* 82–4.

Latham, A. G., Schwarcz, H. P. and Ford, D. C. 1986: The

paleomagnetism and U-Th dating of a Mexican stalagmite, DAS2. *Earth and Planetary Science Letters* 79, 195–207.

Latham, A. G., Schwarcz, H. P., Ford, D. C. and Pearce, W. G. 1979: Palaeomagnetism of stalagmite deposits. *Nature* 280, 383–5.

Li, W. X., Lundberg, J., Dickin, A. P., Ford, D. C., Schwarcz, H. P., McNutt, R. and Williams, D. 1989: High-precision mass-spectrometric uranium-series dating of cave deposits and implications for palaeoclimate studies. *Nature* 339, 534–6.

Lyons, R. G., Bowmaker, G. A. and O'Connor, C. J. 1988: Dependence of accumulated dose in ESR dating on microwave power: a contra-indication to the routine use of low power levels. *Nuclear Tracks and Radiation Measurements* 14 (1–2), 243–51.

Noel, M. 1986: The palaeomagnetism and magnetic fabric of sediments from Peak Cavern, Derbyshire. *Geophysical Journal of the Astrological Society* 84, 445–54.

Noel, M. and Bull, P. A. 1982: The palaeomagnetism of sediments from Clearwater Cave, Mulu, Sarawak. *Cave Science* 9 (2), 134–41.

Pease, P. P., Gomez, B. and Schmidt, V. A. 1994: Magnetostratigraphy of cave sediments, Wyandotte Ridge, Crawford County, Indiana: towards a regional correlation. *Geomorphology* 11, 75–81.

Quinif, Y. 1982: Thermoluminescence: a method for sedimentological studies in caves. *Proc. 8th Int. Congr. Speleo., Kentucky*, 1, 308–13.

Schmidt, V. A. 1982: Magnetostratigraphy of clastic sediments from caves within the Mammoth Cave National Park, Kentucky. *Science* 217, 827.

Smith, B. W., Smart, P. L. and Symons, C. R. 1985: ESR signals in a variety of speleothem calcites and their suitability for dating. *Nuclear Tracks* 10, 837–44.

Williams, P. W. 1987: Geomorphic inheritance and the development of tower karst. *Earth Surface Processes and Landforms* 12, 453–65.

Williams, P. W., Lyons, R. G., Wang, X., Fang, L. and Bao, H. 1986: Interpretation of the palaeomagnetism of cave sediments from a karst tower at Guilin. *Carsologica Sinica* 6, 119–25.

SEVEN

Cave Deposits and Past Climates

Introduction

Estimates of past climatic conditions have long held interest for the natural sciences, in that they provide some explanation for the changing patterns of distribution of plant and animal species on the planet, including our own species. Recently there has been renewed interest in past climate reconstruction as a means of providing analogues for the atmosphere likely to result from global warming as a result of the Greenhouse effect. Most attention in this context has focused on the mid-Holocene (4000–7000 years BP) and the Last Interglacial (c.120000–130000 years BP). Climatic shifts can be inferred from the pollen record by analogy with present plant communities of known climatic tolerances. This indirect method makes many assumptions, not the least of which is that past tolerances reflect the present; on the long timescale this may be questionable. More direct methods were pioneered using the record preserved in deep-sea cores, through stable oxygen isotope analysis of foraminifera. These are still the best record available (Shackleton and Opdyke 1973), though long Greenland or Antarctic ice cores may rival them (Jouzel et al. 1987).

On land, a potentially high resolution record of climatic change has been gained from stable oxygen isotope analysis of calcite speleothems (Harmon et al. 1978b; Gascoyne et al. 1981). A past temperature record now exists for most of the last 120000 years in North America, and fragmentary records have been obtained for areas in Europe and Australasia. The method rests on a few basic assumptions. First, that cave temperatures vary little – less than 1°C – with the changing seasons and that relative humidity is close to 100 per cent. This has been shown to be true by long-term measurements, except near entrances and cave streams, where increased air movement occurs. Most cave air temperatures conform to the mean annual temperature of the landscape above the cave. The second assumption is that the cave calcite was deposited in isotopic equilibrium with the surrounding cave air, which means that evaporation was minimal or preferably absent. This can be tested in the method. The third assumption rests on our knowledge of the fractionation of oxygen isotopes as they are crystallized into speleothem calcite.

Basic Principles and Tests for Reliability

The element oxygen has three stable isotopes: ^{16}O, ^{17}O and ^{18}O. Oxygen-16 is about 500 times more abundant than oxygen-18, but the ratio of these two isotopes varies a little in nature and can be used to estimate past temperatures. The ratio of ^{18}O to ^{16}O is usually measured on gas in a stable isotope mass spectrometer. The value is generally measured as the difference between the sample ratio and that of a standard: for water, this is standard mean oceanic water (SMOW), while for calcite the standard is a fossil belemnite, PDB. This method eliminates systematic errors inherent in measuring small concentrations. It is calculated as follows:

$$\delta^{18}O\text{‰} = \frac{^{18}O/^{16}O \ sample - \ ^{18}O/^{16}O \ standard}{^{18}O/^{16}O \ standard} \times 1000 \qquad (7.1)$$

This variation occurs as a result of the process of fractionation. In this process, one isotope is transferred preferentially when there is a change of state – for instance from liquid to vapour. For example, when water evaporates from the surface of the ocean, the vapour contains a higher proportion of the light atom ^{16}O than was present in the ocean water. The vapour is thus depleted in ^{18}O, while the ocean water is enriched in ^{18}O. This enrichment (of the order of 1.1‰ during full glacial conditions) may be preserved in oceanic organisms (such as foraminifera) which use the water for their metabolism, eventually die and become part of the deep-sea sediment. In contrast, ice formed from the vapour is depleted in ^{18}O. These two related processes are known as the *ice volume effect*, affecting stable isotope ratios at high latitudes or preserved isotopic ratios in fossils or crystals formed during the Last Glacial.

Fractionation also occurs during the process of crystallization. If a molecule containing oxygen crystallizes from a saturated solution, or if calcite or aragonite are extracted biochemically from water, then the calcite or aragonite will be slightly enriched with ^{18}O relative to the water. This enrichment is temperature dependent: it is known as the *crystallization effect*, and its magnitude is – 0.24‰. $°C^{-1}$.

Finally, and most usefully, fractionation depends on temperature during condensation and freezing (see Table 7.1). Evaporation and condensation of water at lower ambient temperatures produce lower values of $\delta^{18}O$ in precipitation; thus summer precipitation is isotopically heavier than winter precipitation, and has values of $\delta^{18}O$ that vary from +0.17 to +0.39‰ $°C^{-1}$ greater (Harmon et al.

Table 7.1   Temperature dependence of oxygen isotope enrichment in calcite (from Bowen 1978)

Temperature (°C)	$^{18}O/^{16}O$ ratio in solution	$^{18}O/^{16}O$ ratio in calcite
0	1:500	1.026:500
25	1:500	1.022:500

1978a). This is known as the *precipitation effect*. The temperature relationship is as follows:

$$T(°C) = 16.9 - 4.38(\delta^{18}O_c - \delta^{18}O_w) + 0.10(\delta^{18}O_c - \delta^{18}O_w)^2 \quad (7.2)$$

where O_c = oxygen in calcite and O_w = oxygen in water. The $d^{18}O$ may be reduced by the ice volume effect in high latitudes or during glacial times, and may be increased by the crystallization effect. These two are usually allowed for before final calculations.

From equation 7.2, to determine absolute temperature it is necessary to know $\delta^{18}O$ of the formation water. This is achieved by measuring the 2H:1H, or D/H ratio (Deuterium/Hydrogen), on small inclusions of water trapped in the calcite and sampled by crushing the calcite in a vacuum vessel (Harmon et al. 1979; Yonge 1981). It has been repeatedly shown that there is an excellent correlation between the D/H ratio and $d^{18}O$ in water. This is known as the meteoric water line:

$$\delta D = (8.17 \pm 0.08)\delta^{18}O + (10.56 \pm 0.64)‰$$
$$(r = 0.997, \text{ standard error} = \pm 3.3\%) \quad (7.3)$$

If the D/H ratio cannot be determined, then relative temperature can be assessed from graphs of $\delta^{18}O$ change with time. To date most studies have presented relative temperatures on account of expense and difficulty in measuring D/H ratios of fluid inclusions.

For reliable stable isotope analysis of cave calcite, the mineral must have been deposited in isotopic equilibrium – that is, no evaporation must interfere with the fractionation. This assumption can be tested by taking multiple samples along a single growth layer (see plate 7.1) in the stalagmite. These samples should yield similar values, indicating that slow loss of CO_2 is the dominant process, rather than evaporation, which would tend to produce different values on the outside of the stalagmite. In addition, there should be no correlation between the $^{18}O/^{16}O$ ratio and the $^{13}C/^{12}C$ ratio along a single growth layer. Such a correlation would again suggest rapid loss of CO_2 by aeration rather than slow random processes. If these requirements can be met, then samples taken along the growth axis of a stalagmite dated by other means can provide a relative or absolute temperate curve for the time covered by the speleothem growth Gascoyne (1992).

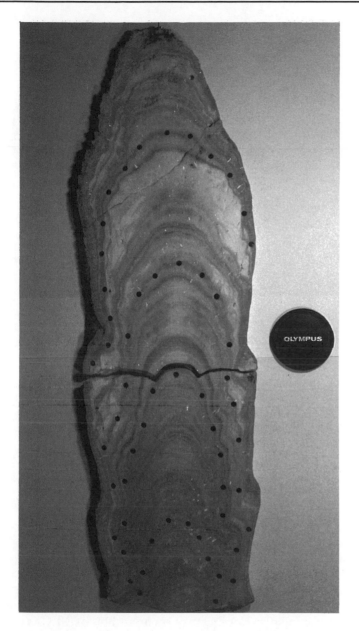

Plate 7.1
*Sectioned stalagmite from
Mimbi Cave, Kimberleys,
showing growth layers and
evidence of three phases of
development. Samples for
stable isotope analysis have
been drilled out along the
growth lines.*

The Last Glacial–Interglacial Temperature Record

One of the best compilations of stable isotope data has been made
by Harmon et al. (1978b) from speleothems collected from Mexico,
Bermuda, West Virginia, Iowa and Alberta (see figure 7.1). Based
on uranium series dates, these stalagmites cover the period from
200 000 years to the present, albeit discontinuously. Several syn-
chronous warming and cooling trends can be seen, and these can be

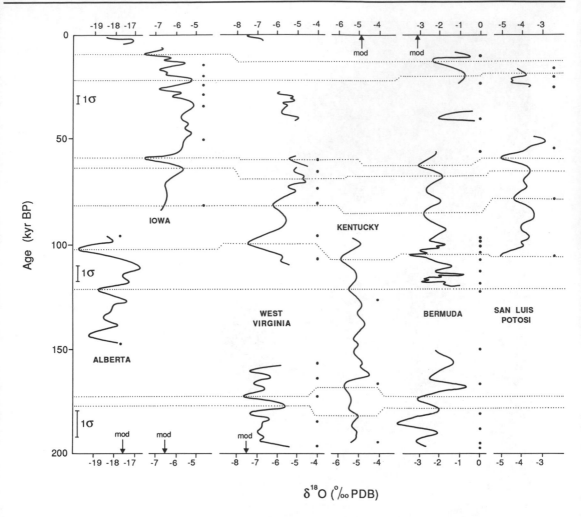

$\delta^{18}O$ (‰ PDB)

Figure 7.1
Record of δ¹⁸O variations in twenty equilibrium speleothems from sites in North America and Bermuda. Solid circles indicate dated points; all peaks are defined by at least three data points. After Harmon et al. (1978b).

compared with the marine isotope record and the Antarctic ice core records. A number of small fluctuations are noticeable between major interglacials and glacials; these may represent periods when the climate was similar to the present. It is also clear from the steep gradients of segments of the isotope curves that climatic change can occur quite rapidly. Cooling may occur at rates of 10°C per thousand years, while warming may be faster, at 15°C per thousand years. Thus the earth-atmosphere system may respond rapidly to changes in insolation or indeed to changes in the insulative properties of the atmosphere, a fact not lost on those scientists studying the Greenhouse Gas phenomenon.

One of the longest continuous records has come from the Flint Ridge–Mammoth Cave system of Kentucky, and this covers the period from 230 ka to 100 ka – that is, the penultimate glacial and the last two interglacials (Harmon et al. 1978a). A section of a long columnar stalagmite was sampled in Davis Hall, a high-level passage with abundant decoration. For this stalagmite, three uranium series dates were obtained, and a continuous timescale was derived assuming constant growth rate between these dated sections. The response of $\delta^{18}O$ to temperature change was determined from fluid inclusion isotope measurements. Major cold periods occur from 215 to 195 and from 160 to 130 ka, while warm periods occur from 180 to 165 and from 125 to 105 ka. These correspond with the marine isotope record, but are better dated (see figure 7.2). Cave temperatures in this region may have differed by a much as 8°C between glacial and interglacial times.

A good palaeotemperature record has been gained by Gascoyne et al. (1981) from Vancouver Island, British Columbia. Speleothem growth was more abundant from 65 to 45 ka, corresponding to an interstadial, and ceased at 28 ka with the onset of the Wisconsin Glaciation. The estimated temperature record shows a steady decline from a mean value of 4°C at 65 ka to 0°C at 28 ka. There is no evidence for warming over that period, nor is there evidence for sudden cooling at 40 to 30 ka, as had been postulated from palynological data. This may be owing to oceanic influences pro-

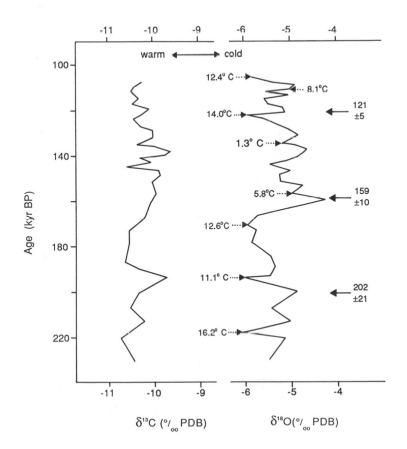

Figure 7.2
$\delta^{18}O_c$ and $\delta^{13}C_c$ variations along the growth axis of a stalagmite from Mammoth Cave, Kentucky. Derived isotopic temperatures and dated points are also shown. After Harmon et al. (1978a).

moting steady temperatures, free from the dramatic fluctuations that might be expected in a continental site.

The Vancouver Island study illustrates a problem encountered in estimating past temperatures from speleothem isotope data. The cooling of cave air during a glacial will result in calcite which is richer in $\delta^{18}O$ than the drip water. This water feeding the stalagmite is ultimately derived from the oceans, which will be poorer in $\delta^{18}O$ during a glacial because of the ice volume effect. These effects therefore work in opposition and may result in underestimation of temperature lowering. Other unknown processes of fractionation may occur between the ocean and the cave drip. It is therefore necessary to look at the isotopic properties of karst water in any study region before undertaking detailed research of stalagmite calcite and fluid inclusions. One place where this has been done is in Tasmania, where a combination of contemporary process measurements and detailed studies of stalagmites has yielded a very fine past temperature record.

Figure 7.3
Monthly precipitation (1–12) against isotopic composition ($\delta^{18}O\%_o$ SMOW) showing temperature effect during summer months. After Goede et al. (1982).

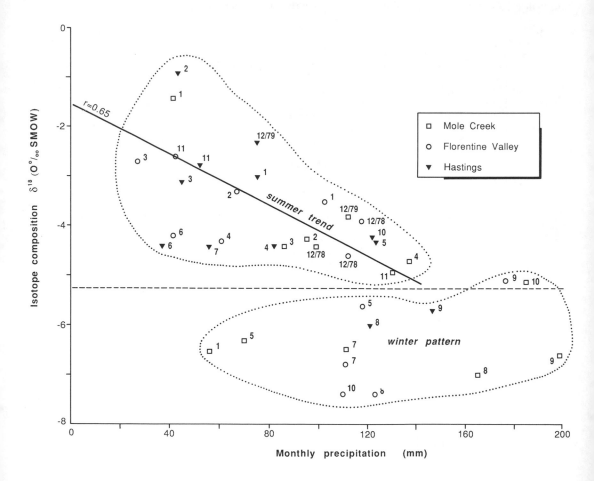

Samples of precipitation and cave drips were collected from three sites (Mole Creek, Florentine Valley and Hastings) along a north–south transect, and analysed for $^{18}O/^{16}O$ (Goede et al. 1982). D/H and $^{13}C/^{12}C$ ratios were also determined for one site and several active stalagmites. All three stations showed a seasonal pattern of variation in the $^{18}O/^{16}O$ ratio, characteristic of the temperature effect (see figure 7.3). There is a clear summer trend distinct from the winter samples, which include some snow. A good meteoric water line was established (see figure 7.4) which was different from that of Dansgaard (1964), normally used as a standard for fluid inclusion studies (Schwarcz and Harmon 1976). Regression analysis between d^{18}O values and mean monthly temperatures yielded the following equations for each site:

Mole Creek: $\delta^{18}O = 0.28T_m - 8.13$ $(r = 0.91)$

Florentine Valley: $\delta^{18}O = 0.39T_m - 8.78$ $(r = 0.76)$

Hastings: $\delta^{18}O = 0.61T_m - 10.13$ $(r = 0.86)$

There is considerable variation in predictor equations between sites, which may relate to regional variation in evaporation. Thus

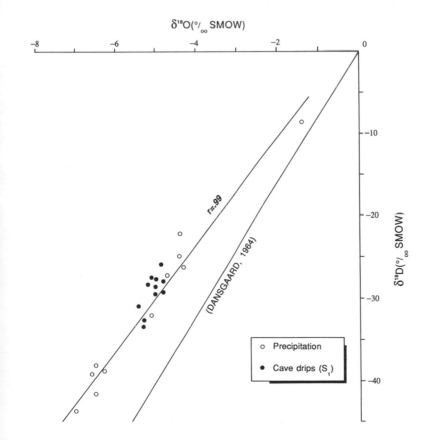

Figure 7.4
Relationship between $\delta^{18}O$ and δD values for precipitation and cave drip at Little Trimmer Cave, Mole Creek. After Goede et al. (1982).

it would be unwise to apply a single equation to estimate palaeotemperatures from isotope data; each site has a unique equation.

The mean isotopic composition of actively forming speleothems was also measured, as shown in table 7.2. Knowing these values, and the mean isotopic composition of cave drips, the temperature of deposition can be calculated. This was done for the mean and seasonal maximum and minimum values for each site. There is very good agreement between these estimates and the observed mean annual cave temperature (see table 7.3), validating the method for Tasmanian speleothems. The poor match for Hastings cave drips is probably owing to very slow percolation of cave water under a caprock, allowing further fractionation to occur.

Armed with these data and equations, Goede then analysed a series of stalagmites from caves in each area (Goede and Hitchman 1984; Goede et al. 1986, 1990). The results in figure 7.5 provide a temperature record for much of the last 100 000 years. Temperatures over that period were in general slightly cooler than at present. The overall decline in temperatures from the more equable climate of the Last Interglacial is recorded in the Little Trimmer Cave and Francombe Cave stalagmites. The hiatus during the Last Glacial, when groundwater was frozen and speleothem deposition largely ceased, is also seen. Temperatures from 100 to 50 ka may need correction for ice volume effects. In the early Holocene temperatures were a little warmer; this can also be inferred from the regional pollen record. The major problem with this and most other speleothem records is the breaks because of cessation of growth or

Table 7.2 Oxygen and carbon isotopic composition of actively forming stalagmites in three Tasmanian karsts (from Goede et al. 1982)

Site	$\delta^{18}O‰$ PDB	$\delta^{13}C‰$ PDB
Mole Creek	−3.65	−11.23
Florentine Valley	−3.77	−6.77
Hastings	−3.67	−9.90

Table 7.3 Relationships between observed and oxygen isotope estimated temperature ranges for three Tasmanian karsts (from Goede et al. 1982)

Site	Observed mean annual temp. (°C)	Calculated minimum temp. (°C)	Calculated mean temp. (°C)	Calculated maximum temp. (°C)
Mole Creek	9.5	9.4	10.0	10.6
Florentine Valley	8.3	8.6	9.3	10.3
Hastings	9.4	4.6	6.8	8.1
Hastings (rainfall)	9.4	9.3	11.2	12.6

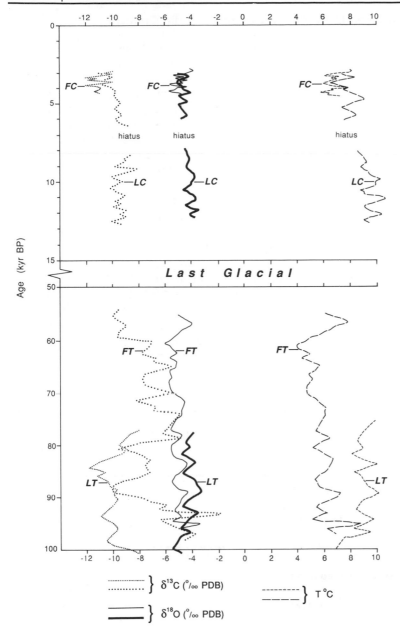

Figure 7.5
Temporal variations in the values of δ¹⁸O, δ¹³C and δD measured along the growth axes of stalagmites from Tasmania. Estimates of past temperature are minimum values. After Goede and Hitchman (1984); Goede et al. (1986, 1990).

resolution of calcite. One of the best long continental climate records from calcite comes from Devils Hole, Nevada (Winograd et al., 1988). A subaqueous deposit has provided a very well dated oxygen isotope record covering the middle to late Pleistocene, 50 to 310 ka. This closely resembles the marine and Antarctic ice core records in its fluctuations, but the chronology suggests that the last interglacial stage began earlier than indicated by the ice core record.

Carbon Isotopes and Environmental Change

Although the $^{13}C/^{12}C$ ratio has been widely used as an index of evaporation in stable oxygen isotope studies and in correction of radiocarbon dates, the $\delta^{13}C$ values have not been widely used for their own palaeoenvironmental content. A measure of relative evaporation can be gained by considering the correlation of $\delta^{13}C$ and $\delta^{18}O$ along the stalagmite growth axis: sections with correlation indicate evaporation, uncorrelated sections indicate more mesic conditions. There are significant differences between the $\delta^{13}C$ ranges of C3 plants (trees and temperate grasses) and C4 plants (tropical grasses and some weeds). These ranges are shown in figure 7.6, along with those of other materials likely to contribute carbon to speleothems. Since soil organic matter is derived from plant matter, the $\delta^{13}C$ values in feedwater will be derived from the surface vegetation above the cave as well as the passage of water through the bedrock. The major exception to this is where hydrothermal leaching of limestone enriches calcite in $\delta^{13}C$ (Bakalowicz et al. 1987). According to Ingraham et al. (1990) more water is lost by leakage from cave pools than from evaporation, and the rate of carbon isotope fractionation in cave pools is limited by the exchange of ^{13}C between water and vapour. The carbon isotope composition of cave deposits has been modelled by Dulinski and Rosanski (1990) with a good degree of success. Thus the system is relatively well understood, and a significant change in $\delta^{13}C$ values

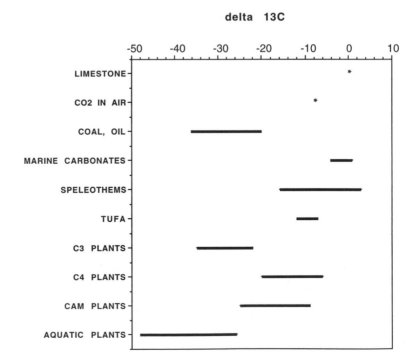

Figure 7.6
Range of $\delta^{13}C$ values for different materials found in the natural environment.

of cave calcite over time may reflect a major change in vegetation. An enrichment of calcite $\delta^{13}C$ from values around -24 per cent to -15 per cent would imply a change from C3 to C4 plants, possibly because of warming. Studies of this kind have been carried out on soil profiles and on tufas (Marcenko et al. 1989), but there is considerable potential to apply this analysis to speleothems.

Other organic compounds such as humic and fulvic acids, and terpenes, have been isolated from calcite speleothems (Caldwell et al. 1982; Lauritzen et al. 1986). Much of the pigmenting material in stalagmite and flowstone is organic material, though there are clear cases where the colour is owing to inorganic trace elements (White 1981; White and Brennan 1990). Spate and Ward (1980) have shown that the black speleothems at Yarrangobilly, New South Wales, are because of finely divided organic carbon, not manganese as hitherto thought. Similarly, the black to dark brown calcite formations of the Nullarbor Plain may be on account of organic matter, though here the colour is probably owing to iron, manganese and organic matter complexing with clays. Many calcite speleothems exibit luminescence in the wavelength range 400 to 500 nm. The brightest luminescence is produced by fulvic acid, with several well defined emission spectral peaks. Changes in the emission spectrum reflect the continuum of molecular structures between humic and fulvic acid. It should be possible to examine the emission spectra along a growth axis and gain information about the mix of humic material that has been captured by speleothem calcite over time. This could be calibrated against the humic acid emission spectra of contemporary vegetation types, and thus provide a long-term record of environmental change. This will be one of the challenges for karst research in the twenty-first century.

Stalagmite Fluorescence and Sunspot Cycles

A radical new technique promises to give us fresh evidence on variations on the solar sunspot activtiy which affects much of the climate and life on earth. This has been developed by the Bulgarian scientist Y. Y. Shopov in collaboration with cave scientists from McMaster University in Canada. As we have seen, calcite stalagmites and flowstones contain isotopic records of palaeoclimate that can be dated by uranium series and other methods. However, these analyses are both time consuming and expensive. Is there an alternative way to gain proxy records of climatic variation?

When stimulated by ultraviolet light (200 to 340 nm wavelength), many speleothems are strongly luminescent. This can be seen if an electronic flashgun is fired close to a stalagmite; there will be an afterglow which persists for a few seconds. Both Shopov (1987) and White and Brennan (1990) showed that this luminescence was owing to calcium complexes with humic and fulvic acids, plus some organic esters. Luminescence spectroscopy must be applied to distinguish this from effects owing to trace elements. Many speleothems display finely banded patterns that are parallel to the visible colour banding; Shopov (1987) showed that these displayed 11- and 22-year cycles and postulated that they were tied to sunspot cycles, though the mechanism remained unclear.

More recent work by Shopov et al. (1994) has employed radiocarbon and Thermal Ionization Mass Spectrometry (TIMS) uranium series dating to refine the chronology of deposition of selected speleothems from Coldwater Cave, Iowa, and from Jewel Cave, South Dakota. The radiometric dating shows that speleothem banding seems to occur on two timescales: fine annual banding of 1 to 100 μm, and long-term banding from 10 to 1000 μm. Annual banding may contain variations in luminescence which may be similar to lacustrine varves. Annual oscillations are very well resolved in many stalagmites examined, while higher frequency components in the banding may represent minor changes in drip rates controlled by individual extreme rainfall events. Over longer timescales, changes in the luminescence reflect shifts in drip rate

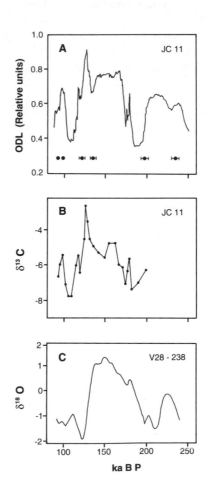

Figure 7.7
a) Optical density of luminescence (ODL) record of upper 47 mm of flowstone sample JC11 fom Jewel Cave, South Dakota. Solid circles are TIMS U-series dates; b) $\delta^{13}C$ record from JC11 sampled parallel to ODL traverse; c) Record of $\delta^{18}O$ of planktonic foraminifera from ocean core V28–238. From Shopov et al. (1994).

and also changes in the vegetation–soil humus complex. For Jewel Cave, South Dakota, a stalagmite luminescence record for the last thousand years shows good correlation with the deep-sea core isotope record (see figure 7.7), and changes in $\delta^{13}C$ may reflect major vegetational change. Jewel Cave lies under pinyon pine woodland, above a transition to C4 grassland; the drop in $\delta^{13}C$ may reflect a lowering of the tree line across the top of the cave. This shift is accompanied by a decrease in luminescence. Thus the technique has potential to detect major shifts in vegetation controlled by climate change or anthropogenic influences. In combination with isotopic analysis, this offers the prospect of a more complete environmental reconstruction than hitherto possible, and may truly unlock the potential of nature's vaults.

References

Bakalowicz, M. J., Ford, D. C., Miller, T. E., Palmer, A. N. and Palmer, M. V. 1987: Thermal genesis of solution caves in the Black Hills, South Dakota. *Bull. Geol. Soc. Am.* 99, 729–38.

Bowen, D. Q. 1978: *Quaternary Geology: a Stratigraphic Framework for Multidisciplinary Work.* Oxford: Pergamon Press, 221 pp.

Caldwell, J. R., Davey, A. G., Jennings, J. N. and Spate, A. P. 1982: Colour in some Nullarbor Plain speleothems. *Helictite* 20 (1), 3–10.

Dansgaard, W. F. 1964: Stable isotopes in precipitation. *Tellus* 16, 436–49.

Dulinski, M. and Rosanski, K. 1990: Formation of $^{13}C/^{12}C$ isotope ratios in speleothems: a semi-dynamic model. *Radiocarbon* 32 (1), 7–16.

Gascoyne, M. 1992: Palaeoclimate determination from cave calcite deposits. *Quat. Science Reviews* 11, 609–32.

Gascoyne, M., Schwarcz, H. P. and Ford, D. C. 1981: Late Pleistocene chronology and paleoclimate of Vancouver Island determined from cave deposits. *Can. J. Earth Sci.* 18, 1643–52.

Goede, A., Green, D. C. and Harmon, R. S. 1982: Isotopic composition of precipitation, cave drips and actively forming speleothems at three Tasmanian cave sites. *Helictite* 20 (1), 17–27.

Goede, A., Green, D. C. and Harmon, R. S. 1986: Late Pleistocene paleotemperature record from a Tasmanian speleothem. *Aust. J. Earth Sci.* 33, 333–42.

Goede, A. and Hitchman, M. A. 1984: Late Quaternary climatic change: evidence from a Tasmanian speleothem. In J. C. Vogel (ed.) *Late Cainozoic Palaeoclimates of the Southern Hemisphere.* Rotterdam: A. A. Balkema.

Goede, A., Veeh, H. H. and Ayliffe, L. K. 1990: Late Quaternary palaeotemperature records for two Tasmanian speleothems. *Aust. J. Earth Sci.* 37, 267–78.

Harmon, R. S., Schwarcz, H. P. and Ford, D. C. 1978a: Stable isotope geochemistry of speleothems and cave waters from the Flint Ridge–Mammoth Cave system, Kentucky: implications for terrestrial climate change during the period 230 000 to 100 000 years BP. *J. Geol.* 86, 373.

Harmon, R. S., Thompson, P., Schwarcz, H. P. and Ford, D. C. 1978b: Late Pleistocene paleoclimates of North America as inferred from stable isotope studies of speleothems. *Quaternary Research* 9, 54–70.

Harmon, R. S., Schwarcz, H. P. and O'Neil, J. R. 1979: D/H ratios in speleothem fluid inclusions: a guide to variations in the isotopic composition of meteoric precipitation. *Earth and Planetary Science Letters* 42, 254–66.

Ingraham, N. L., Chapman, J. B. and Hess, J. W. 1990: Stable isotopes in cave pool systems, Carlsbad Caverns, New Mexico. *Chemical Geology (Isotope Geosciences)* 86, 65–74.

Jouzel, J., Lorius, C., Petit, J. R., Genthon, C., Barkov, N. I., Kotlyakov, V. M. and Petrov, V. M. 1987: Vostok ice core: a continuous isotope temperature record over the last climatic cycle (160000 years). *Nature* 329, 403–7.

Lauritzen, S. E., Ford, D. C. and Schwartz, H. P. 1986: Humic substances in speleothem matrix: paleoclimatic significance. *Proc. 9th Int. Congr. Speleo., Barcelona*, 77–9.

Marcenko, E., Srdoc, D., Golubic, S., Pezdic, J. and Head, M. J. 1989: Carbon uptake in aquatic plants as deduced from their natural ^{13}C and ^{14}C content. *Radiocarbon* 31 (3), 785–94.

Schwarcz, H. P. and Harmon, R. S. 1976: Stable isotope studies of fluid inclusions in speleothems and their paleoclimatic significance. *Geochimica et Cosmochimica Acta* 40, 657–65.

Shackleton, N. J. and Opdyke, N. D. 1973: Oxygen isotope and palaeomagnetic stratigraphy of Equatorial Pacific Core V28–238: oxygen isotope temperatures and ice volumes on a 10^5 year and 10^6 year scale. *J. Quaternary Research* 3 (1), 39–55.

Shopov, Y. Y. 1987: Laser luminescent microzone analysis: a new method for investigation of the alterations of climate and solar activity during the Quaternary. In T. Kiknadze (ed.) *Problems of Karst Study in Mountainous Countries*. Tbilisi, Georgia: Metsniereba, 228–32.

Shopov, Y. Y., Ford, D. C. and Schwarcz, H. P. 1994: Luminescent microbanding in speleothems: high resolution chronology and paleoclimate. *Geology* 22, 407–10.

Spate, A. P. and Ward, J. J. 1980: Preliminary note on the black speleothems at Jersey Cave, Yarrangobilly, NSW. *Proceedings 12th Biennial Conference, Australian Speleological Federation*, 15–16.

White, W. B. 1981: Reflectance spectra and colour in speleothems. *NSS Bull.* 40, 20–6.

White, W. B. and Brennan, E. S. 1990: Luminescence of speleothems due to fulvic acid and other activators. *Proc. 10th Int. Congr. Speleo., Budapest*, 212–14.

Winograd, I. J., Szabo, B. J., Coplen, T. B. and Riggs, A. C. 1988: A 250000 year climatic record from Great Basin vein calcite: implications for Milankovitch theory. *Science* 242, 1275–80.

Yonge, C. J. 1981: Fluid inclusions in speleothems as paleoclimate indicators. *Proc. 8th Int. Congr. Speleo., Kentucky*, 301–4.

EIGHT

Cave Ecology

Introduction

Caves have attracted the interest of biologists and ecologists for over a hundred years (Schiner 1854; Vandel 1965). Physiological and evolutionary adaptations to cave living occur in many phyla of the animal kingdom, and also in some flowering plants and fungi. The principal changes are a loss of pigmentation, partial or total loss of eyes, extension of sensory hairs or antennae, and changes in body part proportions (Culver 1982). This often produces somewhat bizarre looking animals which have thus attracted attention despite their cryptic habits. These adaptations are broadly classified through a scheme of cave life – trogloxene, troglophile, troglobite – which relates to the proportion of its life that an organism spends in the truly dark zone. In this chapter the basic characteristics of the cave ecosystem and its constituent trophic levels are described. This relates back to the basic concepts of mass and energy flow in the karst system. The role of external energy sources is critical for cave life. Cave biota are dependent on periodic inputs of nutrients, usually swept in by floods. They are also severely disadvantaged by quite minor disturbances. Thus they have low resilience in the face of a change to the cave ecosystem. The spectrum of impacts on caves have serious consequences for their biology and ecology, and adequately conserving cave biota is a major challenge for protected area management.

Life Zones within Caves

The entrance zone of caves is a highly variable habitat organized along a gradient of decreasing light and increasing humidity away from the entrance (Howarth 1983). Around the entrance overhang the environment differs little from the surface, though more shelter is available and flows of mass and energy may be more focused. Light and warmth may be available for longer periods of time. The presence of collapse blocks and relatively coarse sediments may provide a wide range of interstitial spaces for colonization. Greater amounts of organic matter, higher plant production and a greater diversity of potential food for predacious animals may be present. Many of the organisms in this zone are not obligatory cave dwellers, and there may be many accidental visitors here, which have either been washed in or have strayed (see plate 8.1).

In the twilight zone there is still some light, but a more stable temperature regime accompanied by higher humidity. There is less organic matter, but fungi may be important in slow nutrient release by decomposition. Organisms start to show specific adaptations to cave dwelling, but there are many accidental visitors in this zone as well as species which feed outside the cave.

The deep, truly dark zone of an individual cave is a rigorous to harsh environment for most surface- or soil-dwelling organisms. Within a network of passages of varying sizes conditions are perpetually dark (except in tourist caves or frequently visited wild caves), with relatively constant temperatures close to the surface mean annual temperature. The substrate in many middle- to high-latitude caves is perpetually wet and the atmosphere close to saturated with water vapour, often at levels beyond the tolerances of terrestrial arthropods (Poulson and White 1969). Carbon dioxide concentrations may be high, and the seasonal cues – fluctuations in light and temperature – available to surface-dwelling organisms are lacking.

Temperatures are usually constant in caves on account of the buffering or insulating effects of cave walls and roofs and reduced air circulation. A cold trap effect may, however, be present in caves

Plate 8.1
The upstream
entrance of Deer
Cave, Mulu, Sarawak,
provides an energy
and light gradient
from the surface
rainforest to the
perpetual dark of the
cave. Organic matter
is washed into the
cave on a daily basis.

which descend steeply from the entrance, resulting in local drying on rock surfaces. Conversely, blind, upwards trending passages may trap warmer air and provide suitable habitats for bats, with static, saturated air. But most cave atmospheres conform to the mean annual temperature, unless they lie close to the surface and are subject to local air circulation. Air movement may be seasonally variable, but there will be pockets of still air in which spiders can weave delicate webs to entrap their prey. Under such stable conditions the approach of prey or potential predators can be detected by sudden air movements (Howarth 1983).

The humidity regime of many caves is more complex. Locally there may be films of water on rock surfaces, pools and active drips. There may be condensation at constrictions where air movement is pronounced. The constant high humidity in the deep parts of caves has led to the morphologies of troglobitic arthropods being similar to those of aquatic arthropods (Howarth and Stone 1990). Higher carbon dioxide levels in the still air of deep caves may result in slowed metabolism as a physiologic response. Figure 8.1 illustrates the changes in key atmospheric parameters and distribution of animals along a section in Bayliss Cave, north Queensland. Both humidity and temperature increase away from the entrance, which is the normal trend in most caves. Carbon dioxide concentrations increase dramatically at the back of the cave, and troglomorphic species increase in this deep zone with foul air. The saturated air and high CO_2 levels (some 200 times ambient concentration) support a highly diverse community of cave-adapted species tolerant of the conditions there, unlike surface and facultative cave species. Thus carbon dioxide concentrations may be significantly higher in caves than on the surface, owing to degassing from speleothems and the decomposition of organic detritus. This may play a role in the adaptation of cave invertebrates. Their adaptation to radiation sources, such as radon gas, remains unstudied.

Many troglobites have fat stores to cope with intermittent food supplies. Although food sources may be very restricted, this is not always so in modified cave systems. But for many sites the filtering action of cave passages may at the very least mean that inputs of food are very widely spaced in time. In such caves periodic floods may be crucial for the introduction of energy reserves to maintain cave life.

The Cave as a Habitat

Biologists have traditionally regarded caves as finite environments whose boundaries are well defined (Barr 1968) and within which the resident flora and fauna can be censused, their trophic relationships defined and the flows of mass and energy measured. Thus caves have been seen as ideal natural laboratories in which to test ideas about processes of adaptation, the structure and function of

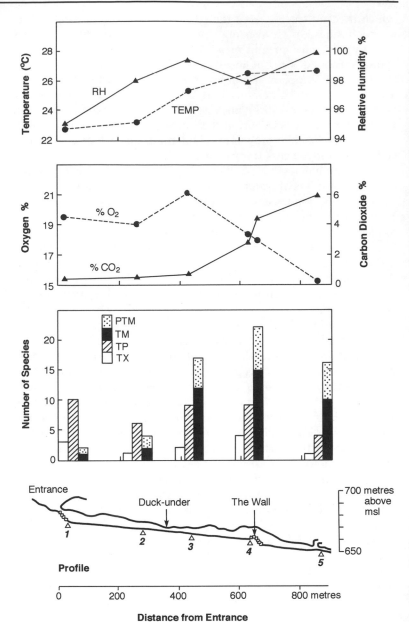

Figure 8.1
*Atmospheric environments
and animal distributions in
Bayliss Cave, north
Queensland, Australia.
Bottom: profile view of
Bayliss Cave with sampling
points. On the bar graph,
PTM = partly
troglomorphic species; TM
= troglomorphic species;
TP = troglophilic species;
TX = trogloxene species.
From Howarth and Stone
(1990).*

simple ecosystems, and the reaction of ecosystems and their components to induced changes. Recent research (Holsinger 1988;
Howarth 1991; Humphreys 1993) suggests, however, that the
boundaries are not as finite as previously imagined. Speleologists
can enter drained and some flooded cave conduits and examine
their fauna and flora. It is rarely possible to examine the extensive
network of fine cracks and the interstitial spaces in sediments which

are normally associated with the conduits. These spaces represent a further habitat for cave organisms which have their own special characteristics and may connect with other conduits, groundwater bodies and the surface environment. There is thus the possibility that caves are not closed environments, and that unexpected interchanges of biota, mass and energy may occur in areas which are at present difficult to sample. This has especial relevance for the survival of small populations of organisms which have become locally extinct in a conduit as a result of some imposed changes in the habitat, such as pollution or increased sedimentation.

The most obvious change in the cave habitat as we move further inside is the gradual reduction in the intensity of light. Depending on the architecture of the passage, the total extinction of daylight may occur from a few metres to tens of metres into the cave. The absence of light deprives cave biota of one major source of energy, and means that the production of biomass by the photosynthesis of green plants is limited to the entrance zone. Thus the base of the ecosystem structure – the primary producer – in most terrestrial and aquatic environments is largely lacking in caves. This is essentially true for nearly all caves, though a type of lichen comprised of gram-positive bacteria and blue-green algal cells coats cave walls in some tropical caves (Whitten et al. 1988). This colony has a high density of photosynthetic membranes (thylakoids) and may be able to use very low light levels. An exception also occurs when seedlings fall or are washed into caves, and produce sprouts from their stored starch reserves. Thus pale or white seedlings may be found several hundred metres inside caves, and may provide an extra energy source for cave biota.

This reduction in primary production by photosynthesis has important consequences for the types of organisms which live inside caves and their lifestyles. Most are by necessity detritivores or carnivores. The only common primary producers in such an ecosystem are bacteria and fungi. Bacteria may be divided into heterotrophs, breaking down organic debris and gaining nutrients and energy, or chemautotrophs, reducing mineral compounds such as iron sulphates and thereby gaining energy and nutrients (Dyson and James 1981). Irrespective of their mode of energy production, their total productivity under natural conditions is likely to be low. Their productivity is enhanced slightly by the commonly alkaline nature of cave sediments and waters, but low temperatures limit their metabolic processes. Fungi slowly break down organic debris and provide a further source of food for detritivores, but again their productivity is limited. This limited productivity means that the total biomass supportable in any cave is circumscribed, and thus cave biota are never numerous under unimpacted conditions.

All cave dwellers are largely dependent on food sources brought into the cave from the surface environment. The accession of food and energy from external sources becomes critical for the survival of viable populations of organisms comprising the ecology (see

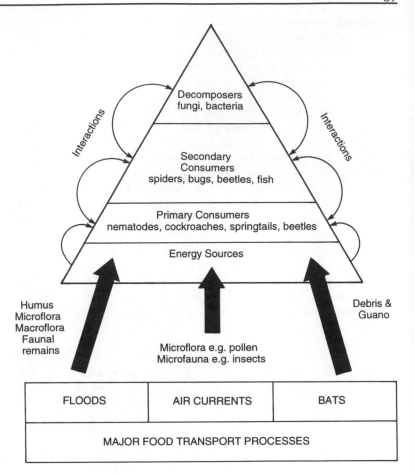

Figure 8.2
*Energy sources, trophic
levels and principal
organisms in a cave
(adapted from Biswas
1992). There are many
complex interactions
between trophic levels
which can only be hinted
at here.*

figure 8.2). The principal source is organic debris washed into the
cave by running water, either as percolation water or as discrete
cave streams. This material may be fine humus, which is readily
utilized by cave biota, or coarser debris (twigs, leaves and
branches), which must first be broken down by bacteria and fungi
to be useable. Sites within the cave that are rarely flooded may
therefore be expected to be depauperate in fauna, while sites along
main streams with a direct connection to the surface environment
may be quite rich in species and in total numbers of organisms
(Poulson 1977). Another significant source of external material is
from air-fall processes below shafts or crack systems open to the
surface. This is especially important for higher-level dry passages
remote from the stream or for caves in dry climates. The penetration of tree roots into cave passages provides a very important
energy source in both tropical and some temperate caves. Some
animals feed directly on the plant roots (see plate 8.2) that penetrate the cave roof, while others feed on detritus that is washed in

during floods or falls in from adjacent daylight holes. As well as the decomposition of bark and stem tissue to produce humus, the exudates from roots provide a source of energy (Howarth 1983). A small amount of organic matter also percolates down through the epikarst and enters the cave via the small roof fissures. Bats and cave-nesting birds also provide food in a number of ways:

- faeces or guano, which has high nutritional value and is fed on by various animals (coprophages) as well as providing a rich nutrient medium for fungi and bacteria
- moulted hair and skin fragments, which provide an additional proteinacous food source
- their progeny may be susceptible to predators and parasites – for example, cave crickets may attack the eggs of swiftlets, and false vampire bats (*Macroderma* spp.) successfully predate on juvenile and adult bats and birds
- they are host to a wide range of ecto- and endo-parasites
- their corpses provide a source of food for animals (necrophages).

There is thus a fairly clear distinction in caves between roof and floor communities (see figure 8.3), each with its own food web but

Plate 8.2
Tree roots provide an extra energy source and habitat in subtropical caves for animals such as this isopod in Yessabah Cave, New South Wales. Photo by Stefan Eberhard.

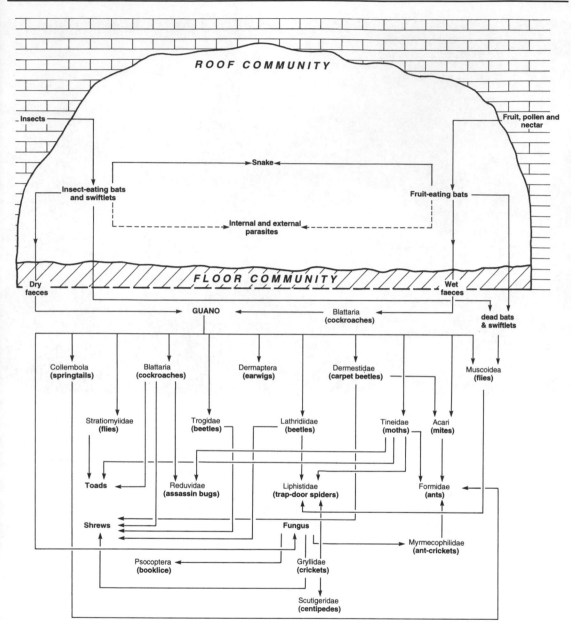

Figure 8.3
Simplified food web from a tropical cave system (adapted from Whitten et al. 1988) showing links between roof and floor communities.

linked by detritus fall. As understanding of the cave community grows, the food web in figure 8.3 tends to become more complex. The floor community is highly diverse, with each group of organisms strongly interlinked with others. The organic layer on the floor of most caves is composed of the waste products and corpses of animals. It is usually rich in nitrogen and phosphorus but often low in carbon because of efficient breakdown of organic matter. On the

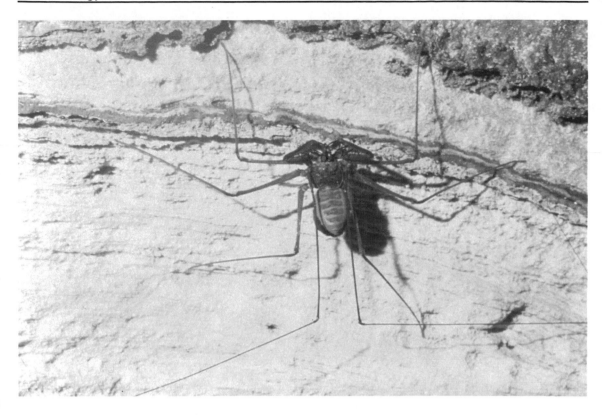

cave floor coprophages and necrophages are dominant, but this distinction may be blurred, with coprophages eating dead animals as well. Coprophages such as woodlice, flies, beetles and Tineid moth larvae will readily exploit the guano of insectivorous bats (Hill 1981). The softer guano of fruit bats will be consumed by cockroaches and some cave crickets. Fungi and some bacteria also digest this food source and in turn are eaten by other organisms such as flies. In addition, intricate parasite or prey–predator relationships become more apparent with detailed research on population ecology. There are a great number of higher order consumers (see plate 8.3) which prey on the animals that consume guano. These include spiders, assassin bugs, ants, crickets and centipedes. Elucidation of the food web can also lead to the construction of population pyramids (see figure 8.4).

The roof community is somewhat simpler and includes bats and swiftlets as well as all those animals that feed on or parasitize them. Bat hawks and pythons are often seen near cave entrances and have been observed feeding on emerging bats at dusk. Bats are host to a wide range of parasites that suck blood or are internal. These include nycterbiid bat flies, ticks, mites, bed bugs and internal roundworms and flukes. Bat parasites show strong adaptation to

Plate 8.3
An Amblypigid predator from Gua Tempurung, Malaya. This 'whip scorpion' is actually related to spiders, and predates on insects, baby birds and frogs. Total length including antennae is 150mm.

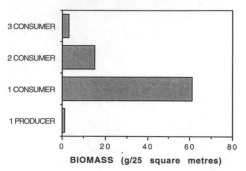

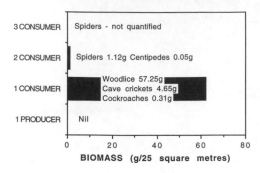

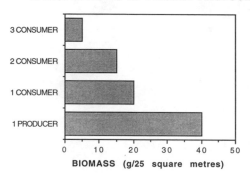

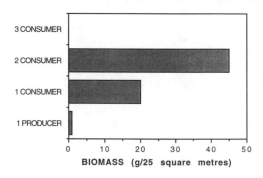

Figure 8.4
Simplified population pyramids for cave ecosystems with a single primary producer, a surface site with a large biomass of primary production, the tropical cave ecosystem of Liang Pengurukan, and a food web based on parasitism. Data from Whitten et al. (1984).

their host and to the cave environment, with body flattening, grasping claws and spines, loss of wings and reduced eyes.

In surface ecosystems the numbers of organisms at the base of a food chain – the producers – are most numerous, while those at the top – the secondary or tertiary consumers – are least numerous. Where parasitism is present, a single host may have a large number of parasites in turn host to a larger number of hyper-parasites (both external and internal). These theoretical structures may also be expressed most usefully in terms of biomass. In surface environments there will be a large biomass of producers and a steadily or exponentially reducing biomass of consumers at each energy or trophic level (see figure 8.4). In cave ecosystems there may be a large biomass of primary consumers but a drastically reduced biomass of secondary consumers. In the tropical cave Liang Pengurukan (Whitten et al. 1984), the total biomass in a 25 m² area of cave floor was 63.38 g (2.5 g m⁻²), of which 90 per cent was contributed by a single species of woodlouse, which accounted for 99 per cent of the animals found. The heaviest animal, a cave cricket, was 900 times heavier than the lightest, the woodlouse. A population pyramid based on numbers alone would overemphasize the woodlouse, but its importance becomes more apparent when the pyramid is constructed from biomass data. In cool climate caves, generally less productive, the differences between trophic levels become even more pronounced.

The frequency and magnitude of energy inputs into the cave ecosystem become very important for the maintenance of populations of organisms. In areas of cold climate with water movement restricted to the spring thaw, biological activity is phased to follow the major influx of water and organic matter, and at other times may be largely dormant. In areas with strongly seasonal precipitation then organisms may have to be adapted to survive desiccation for up to six months, perhaps longer if the climatic variability is high. Cave fauna in tropical areas is less restricted and may be active throughout the year, but reproduction may be phased to reduce competition for resources. Major changes to the frequency of water inputs may have serious consequences for the cave biology, and are common in rural areas where karst water is diverted or overutilized, or if surface changes such as vegetation clearance alter the quantity and quality of percolation water.

Classification of Cave Life and its Function

The most obvious features of the cave environment are the reduced light levels and near-constant temperature regime. Life in total darkness requires that other senses – principally those of touch and smell – become dominant. Thus cave-adapted fauna have greatly enlarged antennae or bristles as well as specialized organs to detect vibration. These adaptations produce some bizarre morphologies in cave animals.

The most widely used classification of cave organisms was developed by Schiner in the late nineteenth century. True cave dwellers or *troglobites* are obliged to live in the deep zone and show significant eye and pigment reduction. These creatures are relatively rare and unable to survive outside the cave environment. The often bizarre cave beetles with absent or residual non-functional eyes and long antennae provide one example of true cave dwellers. Those species which use the deep cave environment but show little eye and pigment reduction are termed *troglophiles* – facultative cave dwellers. They live and breed inside the cave, but on the basis of their morphology it is assumed that they can live on the surface as well, usually in similar dark, humid microhabitats such as the undersides of fallen logs. Many cave crickets, spiders and millipedes fall into this group. Finally those species often found in caves for refuge but which leave to feed are called *trogloxenes*. Bats are a good example of this group (Jefferson 1976). Some other species wander into caves accidentally but cannot survive there.

For aquatic organisms there is a parallel classification. Those highly specialized animals living entirely in the groundwater environment, and absent in surface waters, are called *stygobites*. In contrast *stygophiles* are found in both surface and underground waters without adaptation to subterranean life. *Stygoxenes* are organisms that appear rarely, and almost randomly, in underground waters but are essentially surface dwellers (Ginet and Juberthie 1988; Marmonier et al. 1993).

Adaptations and Modifications to Life in Darkness

The specific adaptations of organisms to life in the cold and dark are:

- loss of pigmentation
- reduction in eyes
- extension of sensory structures
- changes in feeding appendages
- elongation of locomotory spines and claws
- reduced size
- changes to circadian rhythms
- changes to reproductive biology and life cycle (Culver 1982).

These changes in morphology are shown in figure 8.5. As well as dramatic changes in morphology, cave dwellers tend to have altered reproductive strategies involving reduced frequency of breeding and number of larval stages for invertebrates (Ueno 1987). Cave fish tend to have larger eggs and larvae coupled with a longer reproductive cycle. Adaptations to a life in darkness are not necessarily confined to cave dwellers. Troglobites share adaptations with nocturnal animals, in that their non-visual senses tend to be well developed so that they can move, eat and breed in darkness (Nevo 1979). However, their eyes tend to be reduced in contrast to the enlarged eyes or highly responsive retinas of nocturnal animals. Very long appendages, such as the antennae of cave crickets (see plate 8.4) and scutigerid centipedes, are quite normal in troglobites. These antennae can also function as chemoreceptors and may be sensitive to relative humidity (Jones et al. 1992). Echo-location is a well developed trait in bats and some cave-dwelling birds such as swiftlets. Swiftlets click their tongues at low frequencies (1.5 to 5.5 kHz) and the sound echoes are interpreted to produce a 'picture' of the surroundings. This enables them to navigate around cave passages, but is inadequate for hunting their prey of small wasp-like insects, which must be sought in daylight. In contrast insectivorous bat echo-location is highly complex, emitted at high frequencies (20 to 130 kHz) and very accurate, and allows hunting in total darkness for the bats' diet of moths and beetles.

As well as the loss of or reduction in eyes, many troglobite invertebrates have lost cuticle pigments and wings. Some have also developed a thinner cuticle and a more slender body, as they are free of the desiccation of the surface environment. Foot modifications (hooks and bristles) to allow walking on wet or slippery

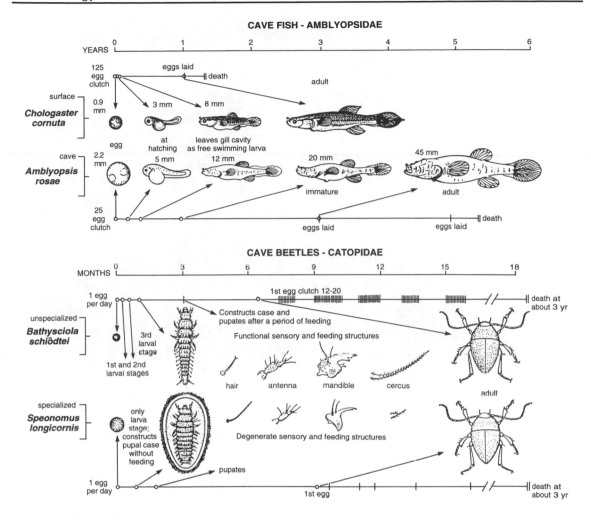

surfaces are common, as are lower metabolic rates. These equip the individual organism to move more securely in the cave and to utilize the available energy sources most effectively.

Most animals have a clearly defined daily cycle of activity. Thus nocturnal species are active at night, diurnal species during the day, and crepuscular species around dawn and dusk. Such cycles are clearly associated with light intensity changes and may be irrelevant to a cave community. There may well be other events that impose a daily rhythm on cave inhabitants. The departure and later return of surface-feeding bats or swiflets causes a change to the rain of urine and faeces from the roof of the cave, a potent energy source for floor-dwelling organisms. Ectoparasites are temporarily deprived of their food source. Thus especially in tropical caves there may be a daily cycle despite the lack of light.

Figure 8.5
A comparison between surface and cave-adapted forms in fish of the family Amblyopsidae and beetles of the family Catopidae. Note changes in gross morphology and size, sensory organs and appendages, and reproductive cycles. Adapted from Poulson and White (1969).

Origin and Dispersal of Cave-Dwelling Animals

The isolation of species in caves may be the result of a variety of
causes, including climatic change, sea-level change and geomorphic
processes. Whatever the cause, isolation promotes evolutionary
changes that make cave faunas of special interest to ecologists and
evolutionary biologists, and pose special problems for conservation
biology. It is clear from recent taxonomic work that many families
and genera of cave-adapted animals have a long evolutionary
history, and have dispersed from centres of origin in the super-
continents Laurasia and Gondwana (Humphreys 1993). Thus cave
fauna in the Cape Range, western Australia, show strong affinities
with those in western India, while cave amphipods in North
America and northern Europe are closely related. It is clear that
many caves are of great age and transcend the Quaternary, the
Tertiary and possibly the Mesozoic. Thus there has been a considerable time – 200 million years or so – for evolutionary adaptation
of cave-dwelling animals to occur.

There are many examples of markedly disjunct distributions
in cave animals. The aquatic amphipod *Stygobromus* – a small
crustacean – is widely distributed in caves of North America, including some several hundred kilometres north of the limit of

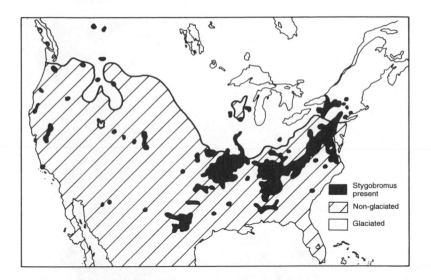

Figure 8.6
Present distribution of the
cave-adapted crustacean
Stygobromus *in relation to*
Pleistocene glacial limits.
Isolated populations have
survived glaciation in ice-
free areas on nunataks or
south-facing valleys. From
Holsinger (1981).

continental ice during the Last Glacial (see figure 8.6; Barr 1973; Holsinger 1981). Many of its habitats are widely separated by non-karstic terrain, the species attaining something like a continuous distribution only in the extensive karst of the Appalachians. Local extinction of populations of this species are now occurring because of water pollution, chiefly from increased nitrate levels owing to herbicide and fertilizer use. Those populations north of the ice limit survived in montane refugia, inhabiting caves in nunataks or sheltered south-facing valleys. The survival of an aquatic species implies that cave streams and groundwater remained unfrozen despite the proximity of ice several kilometres thick (Holsinger and Culver 1988).

There are two main theories of cave colonization and troglobite evolution: 1) the Pleistocene effect theory (Vandel 1965; Barr 1968; Peck 1981); 2) the adaptive shift theory (Howarth 1993). The Pleistocene effect theory has been the most widely accepted model for the evolution of terrestrial troglobites until quite recently. During cold glacial climates, the cooler, wetter conditions south of the continental ice masses of Europe, Asia and America favoured the spread of invertebrates inhabiting both temperate forest ecosystems and caves. With the amelioration of climate, those taxa that survived were those living in caves as the forest ecosystems changed radically. Ultimately geographic and genetic isolation in these cave refugia produced adaptive radiation and the evolution of distinct troglobites (see plate 8.5). In favour of this theory are the close affinities between closed forest and cave taxa (many forest species becoming accidental cave visitors or trogloxenes), the present distributions of taxa in mountain areas separated by deep valleys, and former wider distributions evidenced from the fossil record. Thus the evolution of the troglobitic beetles of the genus

Plate 8.5
The cave-adapted beetle
Idacarabus troglodytes *is*
endemic to the Ida Bay
karst of southern
Tasmania's World
Heritage area. Photo by
Stefan Eberhard.

Pseudanophthalmus, today represented by some 240 species in the eastern United States, may be on account of this mechanism. The surface ancestors of this genus may have spread out from forest habitats in the Allegheny Plateau during cooler glacial periods, and have become a series of isolated populations in the Appalachians and western plateaux such as the Cumberland of Tennessee (Barr and Holsinger 1985). Similarly the cave beetle fauna of southern Australia evolved from a widespread ancestral stock, but the eastern and western populations became isolated by the formation of the arid Nullarbor Plain, an effective barrier to movement (Moore 1964). Cave beetles with affinities to both groups survive on the arid karst today. Overlapping distributions of two troglobitic species may suggest invasion during different glacial cycles, such as that provided by the cave harvestmen *Hickmanoxyomma cristatum* and *H. clarkei* in southern Tasmania (Kiernan and Eberhard 1990), which inhabit the deep and entrance zones of caves.

The second theory, that of adaptive shift, was advanced by Howarth (1993) to explain the origin of troglobites in the Hawaiian lava tubes, but may have much wider applicability. This theory does not rely on climate change; rather it proposes that partially adapted ancestral species moved into cave niches almost continuously. These may have been species out-competed in surface environments. Thus the availability of food is the keystone of this theory, and troglobite evolution has been continual rather than episodic. Caves are buffered against climatic change, and individual taxa may survive for a very long time.

In the seasonally humid tropics, climate change may also have been important, the contrast being wet–dry cycles rather than warm–cool cycles. But the emerging picture of major Pleistocene changes in low latitudes (to which cave studies have made a major contribution) may have more implications for troglobite biogeography than has been realized. There is a very large number of tropical troglobites, and finding a theory which explains their presence is a major challenge. It may well be that no single theory of troglobitic evolution will be valid in all circumstances.

For some trogloxene animals, such as forest cockroaches and centipedes, the cave is an acceptable habitat but only one of their potential living places. But many cave animals are more or less confined to caves, and often to particular cave microhabitats. Caves to them represent islands of biological opportunity in a sea of inhospitable habitats, or barely hospitable habitats inhabited by unhospitable competitors or predators (Howarth 1993). Dispersal can therefore be a serious problem to some cave dwellers. In West Virginia's Greenbrier County, seven blocks of limestone are set among other non-karstic rocks or are divided from each other by rivers – effective barriers to dispersal (Culver et al. 1973). Cave-dwelling animals can move within a block but not between blocks. The number of terrestrial arthropods is correlated with block size, but there is no such correlation for the aquatic arthropods. Presumably they can move in groundwater and some surface waters. Thus the terrestrial animals are vulnerable in the face of habitat change – either climatic or humanly induced – while the aquatic species have more of a chance of survival by dispersal or staying in small refugia.

The network of small fissures and enlarged joints is of great importance for cave-dwelling animals. They are relatively stable, protected habitats where small populations may survive local extinction caused by adverse conditions in a main cave conduit. Many of these fissures are unenterable by people, and thus the recolonization of the major trunk passage by small populations comes as a pleasant surprise to biologists and cave managers. This has happened in Horse Cave, Kentucky, where the removal of pollution sources and some rehabilitation has permitted the resettlement of the main cave passage by fauna which survived in small tributary fissures (Lewis 1993b).

Similarly, the zone of saturated coarse sediments in a cave stream may be an important refuge for aquatic troglobites (see plate 8.6). In surface environments the interstitial spaces in saturated coarse alluvium – the hyporheic zone of stream ecologists – is an important habitat, allowing taxa to survive the drying of stream beds and to escape surface impacts (Allan 1994). The spelean equivalent has

Plate 8.6
*A new species of aquatic
Syncarid (Psammaspides
sp.) from Wellington
Caves, New South Wales.
Such organisms are highly
adapted and are vulnerable
to changed water quality.
Photo by Peter Serov.*

the advantage of near-constant temperature and usually buffered water-level change. It is therefore important that the coarse open texture of many cave stream sediments be maintained, and excessive fine sedimentation be controlled. Where there are connections between cave conduits and gravel-bed surface streams, the hyporheic zone may allow aquatic troglobites to move between caves. Similarly, they may move between caves in deep groundwater as long as oxygenation of the water is adequate.

In common with occurrences in surface environments, cave-adapted fauna may be moved between sites accidentally. It is very easy to move microbiota (bacteria and fungi especially) between caves on boots, rope and clothing. Washing these items between caves is therefore a wise precaution. Aquatic invertebrates may be flushed from cave passages following drainage diversion by cavers or pulse wave release for dye tracing. Bats and birds may act as agents of dispersal for ecto- and endo-parasites; this is especially so in tropical karst regions. Domestic pets such as dogs may also pick up cave entrance invertebrates and move them around.

Threats to Cave Fauna

There are many obvious threats to cave dwelling animals:

- caves are used for legal or illegal dumping of refuse. Although relatively few animals can survive in refuse or garbage, cave faunas are especially sensitive. Increased resource levels may allow non-cave dwelling species to thrive and outcompete the obligate cave species.
- inadequate sewage treatment may pollute groundwater and sediments, leading to local extinction and/or replacement of cave species by surface species.
- cave entrances are vulnerable to closure on account of landfills, housing developments or road construction.
- deforestation or land-use change may affect water flow and sedimentation rates.
- pesticides and herbicides in agricultural area may impact both cave vertebrate and invertebrate populations.
- mining and quarrying remove cave habitats.
- human visitors to caves impact their fauna in many ways.

Many cave species are described from one or two specimens, and extensive searching often reveals only a few individuals. The largest recorded population of large troglobites is 9090 *Orconectes* sp. crayfish in Pless Cave, Indiana (Culver 1986). For most caves, low numbers of endemic species are very vulnerable to pollution by sewage (Illiffe et al. 1984). Serious problems may also arise from pollution by heavy metals and hydrocarbons (Holsinger 1966). The increased bacterial activity in sewage or contaminated water results in loss or great reduction in levels of dissolved oxygen. The animals literally smother in their polluted habitat. Toxins in sewage-contaminated water (endotoxins, heavy metals, phosphates and nitrates) may directly poison cave invertebrates, especially those higher on the food chain. Hamilton-Smith (1970) has discussed the extinction of cave invertebrate faunas by trampling, water pollution and changes in cave microclimates (see plate 8.7). Pesticides and herbicides may have serious effects on cave invertebrates. There are very few studies of nitrate (or other chemical) pollution on caves and karst in Australia. In the porous karst aquifer of southeastern South Australia, Waterhouse (1984) reported background groundwater nitrate-nitrogen levels of 2 ppm and up to 300 ppm at sites grossly contaminated by dairy and piggery wastes. There are also anecdotal accounts of sewage pollution in Tasmania and on the mainland (Kiernan 1988, 8–9); no measured nitrate or phosphate values are given for these cases. Similar levels to the South Australian study were recorded by Quinlan and Ewers (1985) for Horse Cave, Kentucky, where sewage, creamery waste and heavy metal effluent had been detected at 56 springs.

Wheeler et al. (1989) have investigated the impacts of fertilizer and herbicide use on the degraded water quality of the 16 km long Coldwater Cave, Iowa. Average nitrogenous fertilizer application rates on farms nearby were 113 to 170 kg ha^{-1}; atrazine herbicide

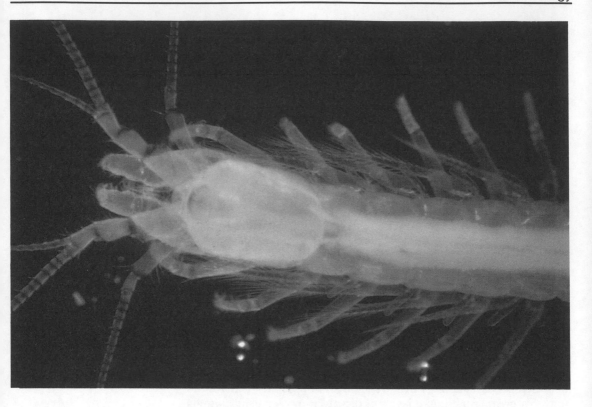

Plate 8.7
A closer view of the anterior of the aquatic Syncarid from Wellington Caves, New South Wales. The animal has strong mandibles, numerous sensory hairs and vestigial eyes. Photo by Peter Serov.

was being applied at 2.2 to 3.3 kg ha^{-1}. Nitrate-nitrogen concentrations in groundwater averaged about 5.8 ppm in the cave stream from 1985 to 1987, with a range of 7.4 to 3.5 ppm. This can be contrasted with a recorded level of 4.1 ppm in 1973–4. Background levels in farm water supply wells were less than 0.1 ppm. Thus the higher recorded values were approaching the Environmental Protection Agency (EPA) maximum standard for drinking water, which is 10 ppm as nitrate-nitrogen (Field 1988). Atrazine was detected in 78 per cent of cave sites, and concentrations ranged from 99 to 1980 ppt (parts per trillion). Thus most of the observed nitrate in cave streams was derived from fertilizer. At these levels observable water-quality decline had occurred (algal blooms, odours).

Vandike (1982) reported on the effects of a fertilizer (ammonium nitrate–urea) spill on karst groundwater in Missouri. Water-quality degradation began at the spring within eight days of the spill: dissolved oxygen levels decreased from the normal 7 to 8 mg L^{-1} to 0.2 mg L^{-1}. Nitrate plus nitrite nitrogen peaked at 4.2 ppm from a background level of 0.05 ppm. Total spill volume was estimated at 24 100 gallons (109 400 L) and the effects on water quality were observable for five months after the accident.

The effects of urban runoff on cave invertebrates was investi-

Table 8.1 Water quality parameters and biodiversity indices for three polluted caves in Tennessee (from Pride et al. 1989)

Site	Dissolved oxygen mgl⁻¹	Nitrate-N ppm	Orthophosphate ppm	Shannon biodiversity index H′
Canal Cave	7.9	0.7	39.4	0.32
Ament Cave	7.5	0.8	55.5	0.25
City Cave	8.9	3.1	56.6	0.49

gated by Pride et al. (1989) for three cave spring sites in Tennessee. The authors measured a range of water-quality parameters and surveyed benthic macroinvertebrate populations. All sites had some history of pollution. The directly storm fed, hydrologically open caves had lower biodiversity indices and significantly higher levels of pollutants. The increased presence of Dipterans and Oligochaetes in both cave streams suggests organic pollution (in this case, from sewage). The cave fed by diffuse recharge had higher biodiversity and better water quality on account of the filtering effect of the karst system; it still contained isopod populations which are regarded as sensitive to pollution. Table 8.1 gives some relevant data from this study. Lowering of dissolved oxygen levels by bacterial uptake will disadvantage aquatic macroinvertebrates, and it is clear that the nutrient concentrations cited are maximal values beyond which major decline of the aquatic fauna would occur.

Increased cave sedimentation may also adversely impact fauna. This may occur in two ways. First, loss of habitat for organisms living or seeking refuge in coarse-textured sediments (sands and gravels) may occur if fine-textured sediments, such as silts and clays, are deposited. Secondly, fine sediments with significant organic matter content may contain bacterial colonies which will also deplete dissolved oxygen, leading to smothering of cave life.

During their long evolutionary history, most troglobites have reduced their capacity to withstand environmental perturbations. Many troglobites cannot withstand desiccation or temperature fluctuations. Cave faunas are thus at especial risk from environmental change.

Conservation of Biological Diversity in Caves

How many species of troglobites are there? We can provide only a crude estimate based on extrapolation from finite, well studied karst areas to continents and then the globe. With the possible exception of birds and mammals, there is no group of organisms on earth for which the total number of species is precisely known. So we are forced to make estimates based on known species diversity from air-filled caves, which may be but a small part of the subterranean habitat. For Virginia, USA, Culver and Holsinger (1992) have listed 160 known species of terrestrial troglobites collected from over 500 caves, a good sample. This covers 15 taxonomic groups, not all of which are described; for amphipods alone, only 29 of the estimated 43 taxa are formally described. Including aquatic troglobites, this total number comes to 549 for Virginia, and perhaps 1000 for the Appalachian karsts. Excluding the diverse fauna of the Hawaiian lava tubes, the US estimate is 6000 species, based on area-species curves. The global estimate, based on proportional area, is about 60 000 species, but the estimates may vary from 50 000 to 100 000. Many of these are known from one cave site only, and many occur in caves subject to pollution. Cave faunas are a fragile resource and the special conservation problems of caves become critical for their biota.

There is thus a great need for surveys of cave faunas at local and regional levels. In several regions of North America and Europe the tireless efforts of individuals have resulted in a good estimate of cave biodiversity and, for certain taxonomic groups, some knowledge of their ecology. But, for large regions of the world, cave faunal survey has been restricted to short visits by scientists. These visits have served only to demonstrate the potential great diversity of cave life, especially in the humid tropics. The studies by Howarth and his co-workers (Howarth 1987; Howarth 1988) have demonstrated the great diversity of lava tube faunas in Hawaii and Australia, and similar species richness is being recorded for tropical limestone caves (Humphreys and Adams 1991).

Given the key role of periodic inputs of food for cave dwellers during floods, there is also a need for seasonal surveys of individual caves and their region. Weinstein and Slaney (1995) have demonstrated that trapping techniques may cause significant variation in species recovery. In particular, the use of wet litter traps, simulating the detritus carried in by floods, may greatly increase the yield of individual sites. Thus our current knowledge of cave organisms is probably only a fraction of the species present. Unfortunately habitat modification may cause their demise before they are sampled, let alone formally described and conserved. Despite the energetic efforts of a growing number of cave biologists, definitive studies of cave ecology at the community level are relatively few. Studies of human impact on cave ecology are limited, and should be a high research priority.

The most widespread classification used for assessing the vulnerability of species is the IUCN classification employed in the Red Data Book (IUCN 1986). There is a hierarchy of categories as follows:

- extinct
- extinct in the wild

- critically endangered
- endangered
- vulnerable
- conservation dependent
- low risk
- data deficient
- not evaluated.

Extinct species are those for which there is no reasonable doubt of their extinction, while *extinct in the wild* species are those which survive only in cultivation, in captivity or as a naturalized population well outside their natural range. A taxon is *critically endangered* when it is facing an extremely high risk of extinction in the wild in the immediate future, as evidenced by severe population decline over the last decade, or when its extent of occurrence is less than 100 km². An *endangered* species is not critically endangered but is facing a very high risk of extinction in the wild in the near future, evidenced by severe population decline of 50 per cent over the last decade, or when its extent of occurrence is less than 5000 km². A *vulnerable* taxon faces a high risk of extinction in the wild in the medium future, evidenced by severe population decline of 50 per cent over the last twenty years, or when its extent of occurrence is less than 20000 km² or its population is fewer than 1000 individuals. *Conservation dependent* taxa must be the focus of a specific conservation programme, the cessation of which would result in its being reclassified into one of the three higher categories. *Low risk* species are those that are close to qualifying for the above, are abundant or are of less concern. *Data deficient* taxa are those for which there are inadequate data to make a meaningful evaluation, but may be listed as threatened when more data become available.

The IUCN Red Data Books provide a useful perspective on the threats to cave and karst faunas, but the listings therein are only a small proportion of cave fauna at risk (see table 8.2). They do, however, provide a good idea of the range of problems facing cave dwellers. The IUCN global review of living resources (World Conservation Monitoring Centre 1992) does not list caves as a habitat type, but has a section devoted to subterranean fishes (WCMC 1992, 121–2, 131–5) which indicates at least 47 species of fish that are either cave adapted or have cave-adapted populations and provides useful data on species, location and adaptations. Population sizes are generally unknown and distributions are localized. It notes that the waters in which these species live or have evolved are the final sumps for water-soluble chemicals used on land, and thus their habitats may be under threat. There is not a comparable review of cave invertebrates, but a planned volume on worldwide cave biology (Hamilton-Smith, pers. comm.) may go some distance to achieving this. Accordingly the Cave Species Specialist Group of IUCN is now compiling information on threatened cave communi-

Table 8.2 Endangered, rare and vulnerable cave species (modified from Culver 1986)

Species	Common name	Known localities	Status	Threatening processes
Troglhyphantes gracilis; *T. similis*; *T. spinipes*	Kocevje underground spiders	Caves near Kocevje, Slovenia	Rare	Industrial development
Adelocosa anops	No-eyed, big-eyed cave spider	Koloa cave and one other lava tube, Kauai, Hawaii	Endangered	Groundwater pollution, withdrawal owing to tourism development
Banksula melones	Melones cave harvestman	16 caves in Tuolumne Co. California	Vulnerable	Flooding from dam
Antrolana lira	Madison Cave isopod	Madison's Cave and Steger's Fissure, Augusta Co., Virginia	Vulnerable	Cave visitors, groundwater pollution
Mexistenasellus wilkensi; *M. parzefalli*	Mexican stenasellids	Cueva del Huisache, San Luis Potosi, Mexico	Rare	Cave visitors, groundwater pollution
Thermosphaeroma thermophilum	Socorro isopod	Sedillo spring, Socorro Co., New Mexico	Endangered	Groundwater pollution; overuse
Grylloblatta chirurgica	Mount St Helens grylloblattid	Lava tubes on Mount St Helens, Washington	Vulnerable	Cave visitors
Speoplatyrhinus poulsoni	Alabama cavefish	Cave in Madison Co., Alabama	Vunerable	Groundwater pollution
Amblyopsis rosae	Ozark cavefish	Caves in Missouri and Arkansas	Rare	Overcollecting
Prietella phreatophila	Mexican blind cavefish	Well in Muzquiz, Coahuila, Mexico	Endangered	Groundwater pollution; overcollecting
Satan eurystomus; *Trogloglanis pattersoni*	Texas blind cavefish	Artesian wells, Bexar Co., Texas	Rare	Groundwater pollution
Typhlomolge rathbuni	Texas blind salamander	Caves and deep wells, Hayes Co., Texas	Endangered	Groundwater pollution; well draining
Proteus anginus	Olm	Caves in former Yugoslavia	Vulnerable	Groundwater pollution; overcollecting
Macroderma gigas	Ghost bat	Caves and mines in arid and tropical Australia	Vulnerable	Human disturbance; quarrying
Myotis sodalis	Social bat	Caves in eastern USA	Vulnerable	Human disturbance
Myotis grisescens	Grey bat	Caves in southeastern USA	Vulnerable	Human disturbance
Plecotus townsendii ingens	Ozark big-eared bat	Caves in Arkansas, Oklahoma, Missouri	Endangered	Human disturbance
Plecotus townsendii virgineanus	Virginia big-eared bat	Caves in Kentucky, Virginia, West Virginia	Endangered	Human disturbance

ties, with the eventual aim of a Red Data Book on caves and their fauna.

The IUCN classification is a useful starting point for consideration of the vulnerability of cave fauna, but by its nature it is species-oriented rather than designed to preserve the whole community and its habitat. In addition, it was designed to cope with species that are areally extensive (such as large herbivores) and whose population dynamics are either well known or amenable to study. The present Red Data Book demonstrates only the minute tip of the iceberg that is the endangered cave fauna of the globe. Despite numerous legal efforts and legislation aimed to give sanctuary to particular cave-dwelling species by designating them as protected fauna, in many cases there is little or no parallel effort adequately to safeguard their habitat or to ensure that ecological studies of species requirements are made to provide a sound basis for conservation. This situation is slowly changing, but in the time taken to enact new legislation many species will become at least locally extinct.

Some major problems with existing legislation are:

- that token gestures such as listing particular species as rare or endangered are futile unless their habitat is also protected and studies of their ecology are conducted
- that scientific collecting is likely to have a minimal effect on their populations compared with ongoing habitat modification or destruction
- that species may not have been selected for legislative protection by any obvious rational scientific procedure related to their scarcity or their biological peculiarity
- that the legislation may at times be hand to enforce because of the difficulty of recognizing certain species by non-specialist personnel.

The Impact of Cavers on Cave Fauna

The general sequence of use of caves by people may be summarized as: local use, discovery by speleologists, increased recreational caving, scientific study, physical and biological resource depletion and, finally, protection of what remains. Cave conservation plans must therefore take account of the impact of visitors on subterranean faunas. The biological impact of repeated cave visits may be far more severe than the physical effects of sediment compaction, flowstone muddying and erosion. Many cave animals have perished under the boots of cavers without the latter being even aware of their existence. The compaction of sediments by repeated visits to a cave removes potential habitat and may result in increased water flows and rill erosion in areas downslope of flowstones and drips. Hamilton-Smith (1970) recorded an important invertebrate site in the unconsolidated floor sediments of a lava tube in western Victoria, Australia; this potentially important study site was wrecked because the landowner encouraged (for a fee) unmanaged wild caving, which reduced the floor to a slick, compacted surface with no fauna at all. Around entrances the increased foot traffic may accelerate the inwashing of fine sediments as well as reducing the carrying capacity of an energy-rich zone of importance to many obligatory or accidental cave dwellers. These impacts are probably more severe for low-energy caves – dry systems with infrequent energy inputs – than for high-energy systems subject to repeated flooding and detritus transfer (Parsons 1990).

The introduction of extra energy sources is one way in which cavers may drastically or subtly alter the cave ecosystem. It is instructive to spread out a plastic sheet and consume a caving snack while seated on it. Around your body will be spread a ring of small portions of the food – chocolate or Mars bar fragments, biscuit crumbs, soup drops – which represent a major local increase in available energy. These items become potential sites for the colonization of fungi and bacteria, or food for cave animals. Even in cold, wet caves, such food fragments will be colonized by a fine fur of fungi in a week or so. The best solution is either to avoid eating in caves or to sit on a plastic sheet which can be carefully brushed afterwards into a bag and carried out. Similarly, human body wastes are potential energy and pollution sources which should either not be deposited inside the cave or *in extremis* deposited in secure containers for removal and disposal off the karst (Tercafs 1992; Poulson 1977). The sudden appearance of *Mondmilch* (moonmilk) in some caves may be related to the introduction of bacteria foreign to the cave by speleologists (Derek Ford, pers. comm.). In both wild and tourist caves there is the potential for these bacteria to be introduced on the fine rain of lint and skin cells normally shed by people every day. Such bacteria may also serve as extra energy sources for cave biota unaccustomed to this rich bounty.

Spent carbide is a potent agent for the demise of cave fauna, especially in low-energy systems where the reaction products may remain for long periods. Calcium carbide reacts to produce acetylene gas and lime, and may contain considerable impurities such as sulphides and some metals. These substances may leach into cave streams or groundwater to pollute them. The unsightly carbide dumps on ledges or in alcoves may remain for periods of several years, and particles may adhere to the appendages of cave animals who walk over them. Although some spent carbide may be flushed by flooding, some always remains in the cave passage, forming unsightly crusts. Carbide dumping in caves is unnecessary, even on expeditions. Spent carbide can be stored in strong plastic jars or bags, or in the innertube 'bombs' used for carbide transport, and then removed entirely from the karst for safe disposal.

The widespread practice of opening entrances by digging or using explosives causes meteorological changes, which may result in the progressive desiccation of rock and sediment surfaces, as well as a reduction in relative humidity. The fumes from explosives also constitute a potential toxin for cave animals. Artificially opened or enlarged entrances should be sealed effectively between visits to minimize changes to air flows and potential desiccation.

The social effects of visitors on bat colonies, especially maternity sites, are a major concern for cave biology (see plate 8.8). There is no question that disturbances as trivial as briefly entering a maternity area with lights can result in decreased survival of young bats

Plate 8.8
*The evening flight of hundreds of thousands of insectivorous wrinkle-lipped bats (*Tadarida plicata*) from Deer Cave, Mulu, Sarawak.*

and possible abandonment of the site (McCracken 1989). Visits to maternity sites early in the breeding season can result in pregnant females moving to other, less favourable sites. The general level of activity by bats is also raised, and this results in greater expenditure of energy and use of fat reserves. This may cause adult mortality in extreme cases and certainly reduced growth of the young bats. Young bats may also become dislodged from the walls and ceiling, falling to the cave floor, where they may perish or be consumed. Clustering of bats in maternity sites has thermoregulatory advantages for the colony. If the size of the colony decreases, by death or by abandonment, then a critical threshold of energy conservation may be reached below which the young cannot be raised successfully.

Hibernating bats are very vulnerable as well. The energy reserves that bats accumulate prior to hibernation are often close to the level necessary for over-winter survival. Disturbance during hibernation can cause bats to arouse prematurely, expending their energy reserves in flight and raising their body temperatures. The bats may go back into torpor after disturbance, but may not have sufficient energy to survive the winter. This may not be apparent to the casual visitor, but a return in spring will often reveal a carpet of dead bats or corpses still hanging on the cave walls.

Human activities outside the cave may also have a negative effect on bat colonies. Vegetation modification by clearing, from controlled burns or from stock grazing, may reduce food sources for bats. There may be other implications for biological conservation in karst regions. In deserts and in tropical areas many plants are pollinated by bats. Their decline or local extinction may affect plant regeneration, especially for cacti. Insectivorous bats are very vulnerable to pesticide poisoning (Clark 1981, 1986), although the reduction of their resource base (insects) and visitor disturbance may be of greater significance. There are several cases of bat colonies in caves badly polluted with toxic chemicals (DDT and PCBs) making population recoveries once the cave was protected (Tuttle 1986).

The ecological effects of tracks and lighting in tourist caves have been well documented, are reasonably well understood and are discussed further in Chapter 9. Developed caves add a new dimension to impacts on cave biology because of the greater physical modification of the cave environment and the addition of extra energy in the forms of light, heat and food.

Unravelling the Secrets of the Carrai Bat Cave

Caves provide homes for a variety of organisms and afford challenges to the ecologists who study them. As we have seen, cave biota are rarely numerous, especially at higher trophic levels, and the cave environment is not an easy one in which to study biology.

Bats roosting in large numbers provide a rich bounty for cave invertebrates. Large supplies of guano build up only in caves which satisfy the microclimatic or social requirements of the bat colony and where there is a low chance of flooding or of active cave streams. Caves or specifically cave chambers which satisfy these requirements are seasonally occupied by many thousands of bats and may be used for over-wintering, mating or maternity sites. Such sites may be used for many years, and the guano heaps in these caves support resident communities of invertebrates of ecological complexity.

One such cave is located in the dense rainforests of northeastern New South Wales. High on a ridge overlooking the deep valley of Stockyard Creek, the Carrai Bat Cave has been the object of study by vertebrate and invertebrate zoologists (Harris, 1970) interested in the dynamics of cave communities. The main outer chamber of the cave is colonized by large numbers of cave crickets, which feed on the leaf litter and humus washed into the cave by slopewash. An inner, enclosed chamber with a domed ceiling is home to between one and three thousand bent-winged bats (*Miniopterus schreibersii*), which roost there from late January to June and again from October to early December each year (see plate 8.9). These regular visits have been repeated with little variation for at least twenty years. The guano heap under their roost is about 150 cm high and 200 to 300 cm across (see figure 8.7). It is inhabited by bacteria, fungi, protozoans, nematodes, mites, flies, moths and

Plate 8.9
Bent-winged bats
(Miniopterus schreibersii)
in flight in a cave in
northern New South
Wales.

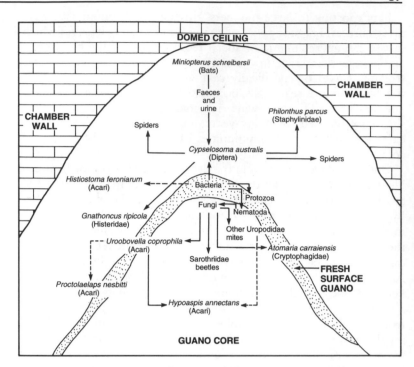

Figure 8.7
Simplified food web of the Carrai Bat Cave guano pile ecosystem. From Harris (1970).

spiders. The bacteria and fungi occur in largest numbers in the surface layers, as do the protozoans and nematodes. Guano mites occur in very large numbers: in particular the small mite *Uroobovella coprophila* (see plate 8.10) has an astounding density of 33 700 000 per square metre! Guano flies (*Cypelosoma australis*) and spiders are found close to the top of the heap. When bats are absent from the cave there is no food source for the guano community. Either the organisms on the heap become inactive, neither growing nor reproducing (as in the case of the guano mite), or their population declines markedly (as in the case of the guano flies). Resting histerid beetle larvae occur thirty centimetres into the heap. Only the staphylinid beetles disappear; presumably they migrate to somewhere else in the cave. The organisms appear to be unable to feed on the core of red, decomposing guano, which is mostly chitinous body parts of moths and beetles.

When the bats arrive in October, fresh dung and urine reinvigorates the community. Microorganisms colonize the guano and their numbers grow rapidly. The guano mites feed, grow and reproduce speedily. Adult flies lay their eggs in the guano, and the mature larvae of the histerid beetles pupate and adults emerge at the surface. The activity of the guano pile is intense for up to two months, and the temperature of the heap increases markedly from 14°C (the cave average annual temperature) to 24°C. In early December the majority of the female bats vacate the cave to give birth at a cave six kilometres away. However, there seems to be

Plate 8.10
In Carrai Bat Cave,
Kemspey, New South
Wales, there is a
high density (33 700 000
animals.m^{-2}) of the
guanophile mite
Urobovella *sp. Tineid*
moths also occur in this
nutrient-rich environment.
Photo by Stefan Eberhard.

enough fresh guano to carry the system until the bats return in late January. From then until June, the numbers of organisms appear stable as a state of equilibrium is attained. This series of changes appear to be a regular annual cycle.

Why do the guano-dwelling organisms not migrate elsewhere in the cave when living gets hard? The central guano core provides not only food, but also a large surface area for colonization and a depth of deposit into which vulnerable larvae can burrow and hide. So those soft-bodied species can be in position for the next input of food in a relatively constant and safe environment. The presence of the core seems necessary for the maintenance of the very large numbers present.

Thus the bat guano community is made up of a small number of species that have evolved to exploit a very specific niche, where the environment is characterized by a regular input of food – the guano – and complex physical conditions where living space is available. Caves are a further demonstration of the ever-changing nature of ecosystems, forming, developing for a while, then collapsing in response to changing energy inputs. Threatening processes that alter these flows of energy include forest clearance in the cave catchment, altered slope hydrology washing too much sediment into the cave, and disturbance to the bat colonies. Fortunately

Carrai Bat Cave is now in the Werrikimbee National Park, and its catchment is largely free from the logging and quarrying that affects other cave sites in the region.

References

Allan, J. D. 1994: *Stream Ecology: Structure and Function of Running Waters*. London: Chapman & Hall, 400 pp.

Barr, T. C. 1968: Cave ecology and the evolution of troglobites. *Evolutionary Biology* 2, 35–102.

Barr, T. C. 1973: Refugees of the Ice Age. *Natural History* 72, 26–35.

Barr, T. C. Jr. and Holsinger, J. R. 1985: Speciation in cave faunas. *Ann. Rev. Ecol. Syst.* 16, 313–37.

Biswas, J. 1992: Kotumsar cave ecosystem: an interaction between geophysical, chemical and biological characteristics. *NSS Bull.* 54, 7–10.

Clark, D. R. 1981: Bats and environmental contaminants: a review. *US Fish & Wildlife Service Spec. Sci. Rept.* no. 235.

Clark, D. R. 1986: Toxicity of methyl parathion to bats: mortality and coordination loss. *Envir. Toxicol. Chem.* 5, 191–5.

Culver, D. C. 1982: *Cave Life: Evolution and Ecology*. Cambridge, MA: Harvard University Press.

Culver, D. C. 1986: Cave faunas. In M. E. Soule (ed.) *Conservation Biology: the Science of Scarcity and Diversity*. Sunderland, MA: Sinauer Associates, 427–43.

Culver, D. C. and Holsinger, J. R. 1992: How many species of troglobites are there? *NSS Bull.* 54, 79–80.

Culver, D. C., Holsinger, J. R. and Barody, R. A. 1973: Towards a predicitive cave biogeography: the Greenbrier Valley as a case study. *Evolution* 27, 689–95.

Dyson, H. J. and James, J. M. 1981: The incidence of iron bacteria in an Australian cave. *Proc. 8th Int. Congr. Speleo., Kentucky* 1, 79–81.

Field, M. S. 1988: *Vulnerability of Karst Aquifers to Chemical Contamination*. Office of Health and Environmental Assessment, Report EPA/600/D-89/008, Environmental Protection Agency, Washington, DC.

Ginet, R. and Juberthie, C. 1988: Le peuplement animal des karsts de France. *Karstologia* 11, 61–71.

Hamilton-Smith, E. 1970: Biological aspects of cave conservation. *J. Sydney Speleo. Soc.* 14 (7), 157–64.

Harris, J. A. 1970: Bat-guano cave environment. *Science* 169, 1342–3.

Hill, S. B. 1981: Ecology of bat guano in Tamana Cave, Trinidad, WI. *Proc. 8th Int. Congr. Speleo.*, Kentucky 1, 243–6.

Holsinger, J. R. 1966: A preliminary study on the effects of organic pollution on Banners Corner Cave, Virginia. *Int. J. Speleo.* 2, 75–89.

Holsinger, J. R. 1981: *Stygobromus canadensis*, a troglobitic amphipod crustacean from Castleguard Cave, with remarks on the concept of cave glacial refugia. *Proc. 8th Int. Congr. Speleo.*, Kentucky 1, 93–5.

Holsinger, J. R. 1988: Troglobites: the evolution of cave-dwelling organisms. *American Scientist* 76, 146–53.

Holsinger, J. R., and Culver, D. C. 1988: The invertebrate cave fauna of Virginia and a part of eastern Tennessee: zoogeography and ecology. *Brimleyana* 14, 1–164.

Howarth, F. G. 1983: Ecology of cave arthropods. *Ann. Rev. Entomol.* 28, 365–89.

Howarth, F. G. 1987: The evolution of non-relictual tropical troglobites. *Int. J. Speleo.* 16, 1–16.

Howarth, F. G. 1988: Environmental ecology of North Queensland caves: or why are there so many troglobites in Australia? *Proc. 17th Bienn. Conf. Aust. Speleo. Fed.*, 76–84.

Howarth, F. G. 1991: Hawaiian cave faunas: macroevolution on young islands. In E. C. Dudley (ed.) *The Unity of Evolutionary Biology.* Vol. I. Portland, OR: Dioscorides Publishers, 285–95.

Howarth, F. G. 1993: High-stress subterranean habitats and evolutionary change in cave-inhabiting arthropods. *American Naturalist* 142, S65–S77.

Howarth, F. G. and Stone, F. D. 1990: Elevated carbon dioxide levels in Bayliss Cave, Australia: implications for the evolution of obligate cave species. *Pacific Science* 44, 207–18.

Humphreys, W. F. 1993: The significance of the subterranean fauna in biogeographical reconstruction: examples from Cape Range peninsula, Western Australia. *Records of the Western Australian Museum Supplement* 45, 165–92.

Humphreys, W. F. and Adams, M. 1991: The subterranean aquatic fauna of the North West Cape peninsula, Western Australia. *Records of the Western Australian Museum* 15, 383–411.

Illiffe, T. M., Jickells, T. D. and Brewer, M. S. 1984: Organic pollution of an inland marine cave from Bermuda. *Marine Environment Research* 12, 173–89.

IUCN 1986: *Red List of Threatened Animals.* Gland, Switzerland: International Union for the Conservation of Nature.

IUCN 1993: *Protecting Nature Regions: Reviews of Protected Areas.* Gland, Switzerland: International Union for the Conservation of Nature, 402 pp.

Jefferson, G. T. 1976: Cave fauna. In T. D. Ford and C. H. C. Cullingford (ed.) *The Science of Speleology.* London: Academic Press.

Jones, R., Culver, D. C. and Kane, T. C. 1992: Are parallel morphologies of cave organisms the result of similar selection pressures? *Evolution* 46, 353–65.

Kiernan, K. 1988: *The Management of Soluble Rock Landscapes: an Australian Perspective.* Sydney: Speleological Research Council, 61 pp.

Kiernan, K. and Eberhard, S. 1990: Karst resources and cave biology. In S. J. Smith and M. R. Banks (ed.) *Tasmanian Wilderness World Heritage Values.* Hobart: Royal Society of Tasmania, 28–37.

Lewis, J. J. 1993b: The effects of cave restoration on some aquatic cave communities in the central Kentucky karst. *Proc. 1991 Cave Management Symposium*, Bowling Green, Kentucky, 346–50.

Marmonier, P., Vervier, P., Gibert, J. and Dole-Olivier, M.-J. 1993: Biodiversity in ground waters. *Trends in Ecology and Evolution* 8, 392–5.

McCracken, G. F. 1989: Cave conservation: special problems of bats. *NSS Bull.* 51, 49–51.

Moore, B. P. 1964: Present day cave beetle fauna in Australia: a pointer to past climate change. *Helictite* 1 (3), 3–9.

Nevo, E. 1979: Adaptive convergence and divergence of subterranean mammals. *Ann. Rev. Ecol. Syst.* 10, 269–308.

Parsons, P. A. 1990: The metabolic cost of multiple environmental stresses: implications for climatic change and conservation. *Trends in Ecology and Evolution* 5, 315–17.

Peck, S. B. 1981: The geological, geographical and environmental setting of cave faunal evolution. *Proc. 8th Int. Congr. Speleo., Kentucky* 2, 501–2.

Poulson, T. L. 1977: Ecological diversity and stability: principles and management. *Proc. National Cave Management Symposium.* Albuquerque, NM: Speleobooks.

Poulson, T. L. and White, W. B. 1969: The cave environment. *Science* 165, 971–81.

Pride, T. E., Ogden, A. E. and Harvey, M. J. 1989: Biology and water quality of caves receiving urban runoff in Cookville, Tennessee, USA. *Proc. 9th. Int. Congr. Speleo., Budapest,* 27–9.

Quinlan, J. F. and Ewers, R. O. 1985: Ground water flow in limestone terranes: strategy, rationale and procedure for reliable, efficient monitoring of ground water quality in karst areas. *National Symposium and Exposition on Aquifer Restoration and Ground Water Monitoring, Proceedings.* Worthington, OH: National Water Well Association, 197–234.

Schiner, J. R. 1854: Fauna der Adelsberger-, Lueger- und Magdalen-Grotte. *Verh. Zool.-Bot. Ges. Wien.* 3, 1–40.

Tercafs, R. 1992: The protection of the subterranean environment: conservation principles and management tools. In A. I. Carmacho (ed.) *The Natural History of Biospeleology.* Madrid: Monografias Museo Nacional de Ciencias Naturales.

Tuttle, M. D. 1986: Endangered gray bat benefits from protection. *Bats* 4 (4), 1–4.

Ueno, S. 1987: The derivation of terrestrial cave animals. *Zoological Science* 4, 593–606.

Vandel, A. 1965: *Biospeleology: the Biology of Cavernicolous Animals.* Oxford: Pergamon Press.

Vandike, J. E. 1982: Hydrogeologic aspects of the November 1981 liquid fertilizer pipeline break on groundwater in the Maramec Spring recharge area, Phelps County, Missouri. *Missouri Speleology* 25, 93–101.

Waterhouse, J. D. 1984: Investigation of pollution of the karstic aquifer of the Mount Gambier area in South Australia. In A. Burger and L. Dubertret (ed.) *Hydrogeology of Karstic Terrains: International Contributions to Hydrogeology* 1 (1), 202–5. Hannover: Heise.

Weinstein, P. and Slaney, D. 1995: Invertebrate faunal survey of Rope Ladder Cave, northern Queensland: a comparative study of sampling methods. *J. Aust. Ent. Soc.* 34, 233–6.

Wheeler, B. J., Alexander, E. C., Adams, R. S. and Huppert, G. N. 1989: Agricultural land use and groundwater quality in the Coldwater Cave groundwater basin, upper Iowa River karst region, USA: part II. In D. Gillieson and D. Ingle Smith (ed.) *Resource Management in Limestone Landscapes: International Perspectives* Special Publication no. 2, Department Geography & Oceanography, University College, Australian Defence Force Academy, 249–60.

Whitten, A. J., Damanink, S. J., Anwar, J. and Hisyam, N. 1984: *The Ecology of Sumatra.* Bogor, Java: Gadjah Mada University Press.

Whitten, A. J., Mustafa, M. and Henderson, G. S. 1988: *The Ecology of Sulawesi.* Bogor, Java: Gadjah Mada University Press.

World Conservation Monitoring Centre (WCMC) 1992: *Global Biodiversity: Status of the Earth's Living Resources,* ed. B. Groombride. London: Chapman & Hall, 585 pp.

NINE

Cave Management

Introduction

Caves have long been of importance to people – for shelter, water supply and food and as objects of veneration and awe. Recent uses of caves include mining of cave formations and guano (especially in the American Civil War), hydroelectricity from cave streams and springs, and as sanatoria. Increasing cave tourism worldwide presents problems on account of the irreversible degradation of cave ecosystems from both point source and dispersed pollution, and alteration of cave microclimates from entrance modifications and visitor numbers. In this context the addition of energy sources (heat from lights and people, lint and dead skin cells) alters the trophic status of developed caverns. Various approaches to cave management have been tried, ranging from the outmoded concept of cave carrying capacity through limits of acceptable change to the more recent systems of cave classification and visitor impact management. There is now a global network of cave scientists addressing these problems, while many governments have recognized the social and economic importance of cave tourism and are allocating resources accordingly. The rehabilitation of caves and karsts after environmental impacts such as mining, forestry and agricultural pollution is a new field in which research and management are closely linked.

History of Cave Use and Exploitation

The earliest direct evidence for human cave use comes from the Peking person (*Homo erectus pekinensis*) site at Zhoukoudian near Beijing (Ren et al. 1981). There the bones and tools of humans are found in the deep cemented fill of a former cave, while nearby much younger fossil human remains have been discovered in a cave sequence of flowstones, scree deposits and ancient hearths. The older Zhoukoudian deposits are dated by palaeomagnetism and uranium series methods to in excess of 700000 years BP and perhaps as much as 1.5 million years BP, while the younger deposits straddle the last glacial cycle. The large cave site of Shanidar in Iraq (Solecki 1972) contains evidence for occupation at *c*.40000 years BP by Neanderthal people (*Homo neanderthalensis*), and there are numerous cave sites of modern people (*Homo sapiens*), such as Cresswell Crags, England (Campbell 1977), Lascaux in the Dordogne (Brunet et al. 1985; Blackwell et al. 1983), Drotsky's Cave in Southern Africa (Burney et al. 1994) and Kutikina Cave in Tasmania (Kiernan et al. 1983), all dating to the onset of the Last Glacial. Superb examples of palaeolithic art can be seen on the walls of limestone caves such as Altamira, Spain; Niaux, Pyrenees; and Lascaux, Dordogne (Andrieux 1983; Brunet et al. 1985; Villar et al. 1986). In Australia the cave paintings of Kakadu and the Kimberley are well known (see plate 9.1). One of the earliest pictures of a cave (at the source of the Tigris) is on an Assyrian bronze panel made around 850 BC. This represents a section of the cave with stalagmites fed by water drops (Hill and Forti 1986, 3).

Caves have long been used for defensive purposes. Many fossil river caves in the Guilin tower karst have partially walled entrances. There are many medieval fortified caves in Switzerland in the Grisons and Vallais – dating to 1160 AD in the case of Marmels Castle, Oberhalbstein, Chur. These small keeps were no doubt safe on account of their difficult access, but were cold and damp refuges. Surprisingly there were cave sanatoria in the USA (see plate 9.2) and Europe where tuberculosis patients were housed in the mistaken belief that the humid air and constant temperature would aid recovery (Lyons 1992). Caves have also been widely used for cheese making and for rope works (see plate 9.3).

The mining of cave guano deposits for fertilizer is a worldwide phenomenon. There is a large array of phosphatic and ammoniacal minerals derived from guano (White 1988, 229). In the USA nitrate minerals were widely mined for the manufacture of saltpetre during the American Civil War (De Paepe and Hill 1981). In Niah Cave, Borneo, cave swiflet guano is mined for fertilizer (Bellwood 1978), while caves at Wellington, Timor, Moore Creek and Mount Etna in Australia were all extensively mined for horticultural fertilizers well into the twentieth century. Caves have long been used as water sources. Many Chinese karst springs were important documented water sources in the Shang dynasty (1600–1100 BC), and the Jinci spring was being used for irrigation in 450 BC (Yuan 1991, 6). Kentucky bourbon from the Jack Daniels distillery still relies partly on cave spring water, though the efficacy of the cave to filter the water may be doubtful. Many Chinese springs are multiple-use sites exploited for hydroelectricity, irrigation and as minor tourist caves – for example, Longgong Cave, Guizhou.

Finally, the growth of cave tourism, from modest beginnings in the late nineteenth century with candle lanterns to today, when fibre optic lights and electric trains are employed, has expanded the

Plate 9.1
Cave paintings by unknown Aboriginal artists at Barralumla Spring, Kimberleys, Australia.

Plate 9.2
*Underground flush toilets
in the Historic Section of
Mammoth Cave,
Kentucky. These elaborate
structures started as huts
for tuberculosis sufferers.*

Plate 9.3
*The remains of Victorian
rope works in the entrance
of Peak Cavern,
Derbyshire, as well as
modern structures
associated with the tourist
cave.*

range of impacts on caves drastically. Cave management is now a
well established science, with leading professionals employed
by government departments and a small number in the private
sector. At a global level the International Union of Speleology,
the International Speleological Heritage Association, the International Geographical Union and the Commission for National
Parks and Protected Areas, IUCN, are all active in cave and karst
management.

Impacts of Visitors and Infrastructure on Tourist Caves

Every year around 20 million people visit tourist caves, with Mammoth Cave, Kentucky, alone receiving over 2 million. There are
some 650 tourist caves with lighting systems worldwide, not counting caves used for 'wild' cave tours, where visitors carry their own
lights. The infrastructure of tourism has a major impact on the cave
system. In Carlsbad Cavern, New Mexico, there are more than
1400 lights, 65 km of electrical wiring, 6 km of walking paths, 70
signs and 56 drains! Surface impacts resulting from the construction of car parking areas, walking tracks, kiosks, toilets, hotels and
motels, and interpretive centres may be added to the direct underground impact.

The range of factors producing environmental impacts on caves
and karst has been identified by Williams (1993, see figure 9.1).
Changes to cave hydrology and atmosphere, to cryptogam growth,
to speleothems and to cave biota are all possible. There is a great
potential for hydrologic change within caves from the construction
of pathways, entrance structures, car parks and toilets. Above a

cave, the surfacing of the land with concrete or bitumen renders it nearly impermeable, in contrast to the high natural permeability of karst. Thus the feedwater for stalactites may be drastically reduced or eliminated. Drains may alter flow patterns and may deliver additional percolation water to certain areas of a cave, causing changes in speleothem deposition. One way to minimize these effects is to use gravel-surfaced car parks or to include infiltration strips and cross-drains in the car park design. Similarly, pathways may need to be hardened for foot traffic, but this should be permeable (gravel, raised walkways, pavers) rather than concrete or bitumen. Toilet facilities may leak into karst fissures or conduits. There are many tourist sites where sewage reticulation or septic tank systems have leaked or overflowed into caves. Today there is a growing trend to use either pump-out toilet systems, where wastes are dispersed as sprays or sludges away from the karst, or composting toilets (Clivis Moltrum, Dowmus), where residues are dehydrated and may be subsequently used for fertilizer.

Within caves, pathways and stairs may alter water flowpaths. Impermeable surfaces made of concrete or steel may deflect the natural water movement away from flowstones or stream channels, causing desiccation of formations or increased sediment transport. The use of permeable steel, wooden or aluminium mesh walkways, frequent drains leading to sediment traps, and small barriers to

Figure 9.1
The range of effects and consequent impacts of human activities on karst and caves. From Williams (1993).

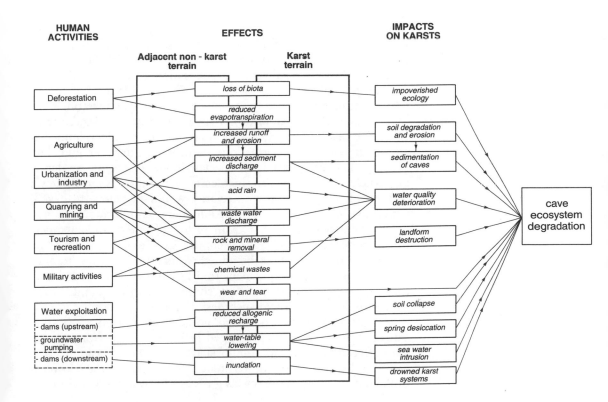

water movement may help to maintain a semblance of the natural flowpaths in a cave.

The main effects of tourists on cave atmospheres are from drying out of cave formations owing to body heat and heat from lighting (increased evaporation), and the effects of increased carbon dioxide levels in dissolving calcite speleothems. A single person in a cave releases heat at a rate between 82 and 116 watts (Villar et al. 1986; $1W = 1J \cdot s^{-1}$), roughly equivalent to a single incandescent light bulb. At Remouchamps Cave, Belgium, a single party of 87 tourists raised cave air temperature by 1.5°C during a five-minute visit (Merenne-Schoumaker 1975). Similarly, a group of 105 tourists in an Italian cavern for seven minutes raised air temperature by 1°C (Cigna 1989). The relaxation time after this rise was about ten minutes. Williams (1975) noted that the daily lighting of the Glow-Worm Cave, Waitomo, could be responsible for a 2°C rise, without the effects of visitor body heat. A 1°C rise in air temperature results in an eightfold increase in water vapour capacity. The passage of tourists through Altamira Cave, Spain, resulted in an increase in air temperature of 2°C, a trebling of the CO_2 content from 0.4 per cent to 1.2 per cent, and a decrease in relative humidity from greater than 90 per cent to 75 per cent. This resulted in widespread flaking of the cave walls, affecting the prehistoric art thereon.

De Freitas and Littlejohn (1987) conducted a detailed study of the climate of this cave, and concluded that the spatial and temporal distribution of air temperature and humidity could be modelled from external air temperature and humidity. In winter there was strong drying of the cave environment as external air entering the cave was warmed. In summer humidity levels in the cave rose substantially and resulted in condensation throughout the cave, despite general warming of the air. They recommended that an upper entrance be closed in winter and that artificial wetting of the cave walls might maintain winter cave humidity. The desiccation of caves as a result of changed airflows is a major problem. Immediate effects are noted on formations, which may dry out and flake. Prolonged changes may lead to a new mode of speleothem deposition, such as cave popcorn. Increased evaporation is most likely to be a problem in winter, when relatively cold air enters the cave by convective currents, is warmed and must equilibrate to a new, greater relative humidity. With an airflow of $1 m^3 s^{-1}$ and a temperature difference of 15°C between the cave and the surface air, $12 g m^{-3} s^{-1}$ of water must be added to maintain relative humidity (Aley 1976). This is equivalent to 120 cave drips a second. Aley (pers. comm.) advocates airflow control structures, similar to refrigerator doors, to avoid such desiccation.

The effects of increased carbon dioxide from visitors' exhalations can be quite serious. A prolonged increase in CO_2 concentration may affect the chemical equilibria of speleothems, leading to re-solution. In New Zealand, Kermode (1974) noted surface lowering of $0.3 mm y^{-1}$ on speleothems as a result of around 500 visitors per day. At Jenolan Caves, Dragovitch and Grose (1990) noted increases in CO_2 concentration of between 2.5 and 4 times as a result of cave tours. The maximum recorded value was 1500 ppm, above the ambient atmospheric level of 320 ppm but well below the human safety limit of 5000 ppm.

The continuous lighting of cave features in tourist caves also provides an opportunity for the colonization and growth of green plants (principally blue-green and filamentous green algae, mosses and ferns) in a concentric zone around the fixture. This is termed *lampenflora* (see plate 9.4), and it has been the subject of a great deal of research and discussion (Hazlinsky 1985). These plants may have a negative visual effect and may actively corrode speleothems. The growth of this 'maladie verte' at the Lascaux prehistoric art site was one factor leading to the latter's closure to the public. Low-wattage lamps, 'cool-lights', periodic ultraviolet irradiation and timed lights can all help to minimize this problem. Cleaning using high-pressure jets and/or strong oxidizing agents (calcium hypochlorite) may be necessary in extreme cases (Slagmolen and Slagmolen 1989). In a Hungarian cave, Rajczy and Buczko (1989) noted 39 species colonizing newly installed lights, with smooth substrate texture being an important factor inhibiting establishment. Significant lampenflora was established on 20 per cent of lamps within a year.

Plate 9.4
Lampenflora coats vertical ribs of stalagmitic formations in Fig Tree Cave, Wombeyan, New South Wales. Photo by Andy Spate.

There are few well documented studies of the incidental effects of tourist cave operations. Most tourist caves have a fine layer of dust on the speleothems (Jablonsky 1990, 1992). This is made up of lint from clothing, dead skin cells, fungal spores, insects and inorganic dust. At Carlsbad Caverns, New Mexico, over 50 kg of lint has been removed manually in the last five years, while at Jenolan Caves, New South Wales, over $0.2\,kg\,m^{-2}$ of lint has been removed from heavily soiled areas. This material has the capability to act as a source of pathogenic bacteria, and may develop an unpleasant odour upon decomposition. It also serves as an extra energy source for cave biota. Visitors' vehicles may contribute a source of dust and other pollutants. James et al. (1990) investigated lead levels in cave spiders' webs at the Grand Arch, Jenolan Caves, and compared levels with nearby non-impacted caves: they found that, although lead levels in the webs were high, they were also high at some other sites, and attributed the high values to dust trapping by webs. The dust caused collapse of webs and may have increased spider mortality. Also at Jenolan, Kiernan (1989) reported minor sewage seepages into the caves from old pipes.

Probably the most extreme example is Hidden River Cave, Kentucky. The underground river served as the water supply for the town of Horse Cave from the 1880s to the 1930s, and there was a booming tourist industry based on the caves of the region. Groundwater pollution on account of domestic sewage from two towns, creamery wastes and industrial wastes (see plate 9.5) destroyed the water source and caused the cave to close in 1943. Blind cave fish in Hidden River Cave became locally extinct, and the only aquatic organisms surviving in the near oxygen-free water were

Plate 9.5
In Water Cave, Wellington, New South Wales, old timbers and fuel drums decompose and corrode slowly, releasing toxic compounds and hydrocarbons into the cave environment. This may adversely affect aquatic invertebrates. Photo by Stefan Eberhard.

bloodworms and sewage bacteria. Since 1989, reduction in sewage as a result of altered municipal practices has led to improved water quality. Troglophilic crayfish, troglobitic isopods and amphipods have returned to the cave stream. Recently the blind fish have been sighted again in the main Hidden River Cave, having recolonized the site from upstream refugia. The cave is now the headquarters of the American Cave Conservation Association, and the recovery of the site is a key feature of their interpretive programme.

Cave cleaning and its impacts

In many tourist caves, formations are cleaned on a regular basis because of the accumulations of dust, lint, inwashed sediments, fungus and algae (lampenflora). A number of approaches have been tried, with high-pressure water jets being the most common method employed (Bonwick and Ellis 1985). In some cases scrubbing, use of surfactants and steam cleaning (Newbould 1976) have also been tried. All these methods can be expected to have some impact on the speleothem surfaces being cleaned.

The impact of cleaning has been assessed by Spate and Moses (1994) for one of the Jenolan Caves, New South Wales. Examination of the unwashed surfaces under ultraviolet light revealed many fibres derived from fabrics washed or bleached with optical brighteners. SEM observations of surfaces before and after cleaning revealed a mean surface lowering of 0.018 mm per wash, with damage to crystal facets being pronounced. On unwashed surfaces individual crystals were obscured by amorphous clay layers thought to be derived from atmospheric dust. Hair and lint were embedded in the clay and in active calcite growths. A single wash at pressures less than 5500 kPa was sufficient to remove these foreign objects, but a second wash was found to expose calcite crystals and damage their facets. Clearly cave cleaning has an undesirable impact; operators should try to limit the number and frequency of washes, and use the minimum number of nozzle passes over a calcite surface. Alternatives to limit particulate entry in tourist caves include plastic mesh walkways at the cave entrance and protective clothing.

Impacts of recreational caving on caves

All visitors to caves have an impact on the cave itself and on its contained biology. We have all been guilty, at some time in our caving careers, of wittingly or unwittingly causing some degradation to a particular cave. Unfortunately the documentation of the impacts of recreational caving is restricted to a few observations, often gained fortuitously as part of another study. Stitt (1977) has outlined the range of internal and external impacts on caves, while Everson et al. (1987) have surveyed recreational impacts on Missouri caves. Specific types of wilderness cave impacts are described by Gamble (1981) and Kiernan (1989). Direct observations of the impact of caving on fauna are restricted to Tercafs (1988) for Belgium, and Carlson (1993), Crawford and Senger (1988) and McCracken (1989) for the United States. This is an area that desperately needs good systematic research, given the fragility of cave ecosystems outlined in the previous chapter. It should be pointed out that many cave scientists also cause significant impacts to the cave environment in the course of their research. These may include the excessive breakage of formations, excessive collection of biota, excavation of pits subsequently left unfilled, permanent marking of study sites or survey stations with inappropriate media (paint, permanent tags, flagging tape), and leaving monitoring infrastructure in the cave. All cave scientists have been guilty of this, to a greater or lesser degree, during their careers. It is now unacceptable practice under most natural or protected area plans of management.

A typical cave is a low-energy environment with essential little input of energy over human activity periods. The entry of a single caver will change the energy regime, albeit slightly, in terms of heat, light and, possibly, nutrients. This impact has little effect on the rock itself unless visitor numbers are large, as we have seen for tourist caves. Cave resources are essentially non-renewable, and these impacts are cumulative, possibly synergistic (Stitt 1977) and irreversible, even when technology and time are brought to bear on a specific problem. For example, a dug entrance could be reblocked, but there will have been changes to the cave atmosphere and possibly its hydrology which may have disadvantaged or eliminated cave fauna. Compaction of sediments may have synergistic effects in that more water is channelled through the cave, sediment may be eroded, infilling of cave pools occurs, cave biota may be smothered by poor aeration, and slumping of sediment banks blocks passages. The impact of a party of ten may be more than twice the impact of a party of five (Stitt 1977). In my youth I was privileged to be one of five people to enter a virgin cave with flat silt floors rich in organic material 10 to 15 cm deep. The cave had a rich invertebrate fauna which was collected by an entomologist on the second trip of twelve people. After six trips about 100 people had seen the cave, the floor was compacted, water flow had been channelized, and the porous silts had been compacted over large areas. The biota was drastically reduced in numbers and species richness in less than a year. Ironically the cave is now under the waters of an irrigation supply dam.

Some specific effects of recreational cavers on caves are:

- carbide dumping and marking of walls
- compaction of sediments and its effects on hydrology and fauna
- erosion of rock surfaces (ladder and rope grooves, direct lowering by foot traffic)

- introduction of energy sources from mud on clothes and food residues
- introduction of faeces and urine leading to water pollution
- entrance and passage enlargement by traffic or digging
- cave vandalism and graffiti.

Many caves which are frequently visited and which are not under any form of access control have suffered from some form of vandalism. This ranges from graffiti to breakage of speleothems and mechanical enlargement of passages. Interestingly, ancient or vintage graffiti become heritage items – for example, eighteenth-century signatures in Baradla and Mammoth Caves; engraved Sung or Ming dynasty lettering in Lung Yin Tung Cave, Guilin, China; and late Victorian era signatures on speleothems beyond the inverted syphon of Murray Cave, Cooleman Plain, Australia. Modern graffitists use paint spray cans; their more durable works can be removed only with strong detergents or solvents.

There is a range of ways of limiting the impacts of visitors on both tourist and wild caves:

- hardening the environment to reduce the impact, by installing tracks, paths and marked routes
- decreasing the demand for cave experiences by restricting the flow of information about caves in the media
- increasing the supply of caves by new discoveries
- exporting the demand to other countries richer in caves
- restricting access to caves by gating and/or permits
- reducing impacts of visitors through education and development of minimal impact codes.

Of all these options, the last is most likely to be productive in the long term, and should involve all identifiable cave users. The development of individual cave management plans by speleologists is one way to ensure responsible conduct (Glasser and Barber 1995) and allows for community consultation in the planning process.

The Radon Risk in Caves

Radioactivity in the environment is a cause of increasing concern to the public, and the gas radon-222 has been highlighted as a potential hazard in houses, mines and caves. ^{222}Ra is released by the radioactive decay of uranium salts weathered from volcanic or plutonic rocks or shales, and it may accumulate in sediments or in areas with poor air circulation. It has a half-life of four days. If this radioactive substance, a daughter of uranium-238, is ingested or inhaled, the alpha and beta radiation associated with it or its daughters may cause serious cell damage, leading to an increased risk of cancer. Radon gas may be absorbed onto particles of dust or onto water droplets, and thus may gain entry to the body. While coarse particles are filtered by nasal hairs and fine particles are exhaled, intermediate-sized particles >10 µm are likely to lodge in the lungs.

The basic unit of radioactivity is the Becquerel (Bq), which is one disintegration per second, and activities are often expressed as Bq m^{-3} for air or Bq kg^{-1} for sediments. Health guidelines are usually expressed in milliSieverts (mSv), which is a measure of the energy absorbed by the body, or as working levels (WL), which are a measure of the radiation dose received related to recommended maximum exposure times; thus we have working level hours (WLH), the dose received by someone exposed to a dose rate of 1 WL for one hour. As a rule, the conversions between these units are as follows:

$$1 \text{ WL} = 0.0735 \text{ mSv} = 3.8 \text{ kBq m}^{-3} \qquad (9.1)$$

Thus a person working for 170 hours in an environment of 1 WL would receive a dose of 12.5 mSv, which is nearly the maximum annual dose for a non-classified radiation worker in the UK (15 mSv). UK ionizing regulations apply above a level of 0.03 WL, while the threshold for government action for domestic radon levels is 0.05 WL. Several cave sites in Britain exceed these thresholds (see table 9.1), especially in areas like the Peak District, where there is mineralization associated with the caves. In Giant's Hole, Derbyshire, an activity of 40 WL means that the annual permissible dose

Table 9.1   British cave radon survey results, 1989–92 (from Gunn et al. 1991; Reaich and Kerr 1991; Hyland and Gunn 1993)

Region	Average[†]	Maximum[†]	Minimum[†]	No. of detectors	No. of caves
Peak District	8716	46080	9	161	12
South Wales	2466	19968	127	168	9
Forest of Dean	1820	4663	182	10	3
Mendips	na	19000	12000	88	1
North Pennines	1079	27136	27	347	18
Portland	454	1024	86	18	5

Potential radiation dose*	Based on averages	Based on maximums			
Peak District	0.36	1.88			
North Wales	0.38	1.16			
South Wales	0.10	0.82			
Forest of Dean	0.07	0.19			
Mendips	0.34	0.44			
North Pennines	0.04	1.11			
Portland	0.02	0.04			

[†] values in Bq m^{-3} radon gas concentration.
* potential doses from a four-hour caving trip, in mSv. A non-classified radiation worker can receive 15 mSv per year while at work under UK legislation.

Table 9.2 Radon measurements in some Australian caves (from Lyons, 1992)

Region	Range (WL)	Mean (WL)	No. of observations
New South Wales			
Jenolan	0.05–0.53	0.33	15
Wombeyan	0.03–0.42	0.14	4
Abercrombie	0.02–0.03	0.03	2
Yarrangobilly	0.23–0.90	0.43	12
Victoria			
Buchan	0.01–0.08	0.06	18
Western Australia	0.02–0.16	0.05	6
Nullarbor (WA & SA)	0.02–0.16	0.05	6

will be reached in one six-hour trip. This does not mean that there is any risk of acute radiation sickness, rather that there is an increased risk of lung cancer in the future as a result of this level of exposure. Caves in Australia (see table 9.2; Lyons 1992) have somewhat lower levels, but there is high variability in the results, and more monitoring needs to be carried out before definitive statements about radon exposure risk can be made (Lyons 1992).

The concentration of radon in cave air will depend on the amount of radon reaching the air and its subsequent dilution. The net amount reaching the air will be the instantaneous sum from all the materials in the cave – rock, sediments, speleothems and water. Dilution factors include the air movement and cave ventilation, the surface area to cave volume ratio, and the loss of radon by plating out on exposed surfaces. Rock usually has low emanation rates, while speleothems have negligible emanation rates. Dry, well sorted sandy sediments will allow ready diffusion of radon and will thus contribute proportionately more radon to the cave atmosphere. Radon concentrations may be higher in areas where there is a higher concentration of uranium salts in rocks or sediments. Caves with deep sediments, especially those which are porous and dry, occurring in smaller or blind passages may have higher concentrations, especially if there is poor air circulation. Caves with extensive speleothems, rock surfaces and water, and with good air circulation, may be less problematic. There may be diurnal or seasonal variations in radon concentration where pressure or temperature gradients drive cave air circulation.

Cave sites where high radon concentrations are suspected should be monitored to take account of these seasonal variations as well as spatial variation within the cave. Staff who spend significant time underground, such as cave guides and electricians, need to have their personal exposure monitored according to prevailing radiation health standards using accepted techniques and instrumentation. Where personal exposure levels are high it may be necessary to modify work practices to limit within-cave time or introduce multi-skilling. In extreme cases it may be necessary to close the cave to visitors seasonally or permanently. Clearly the radon problem is one which affects both tourist caves and recreational cavers, and it is one where levels need to be monitored, work practices amended, and cave ventilation installed if necessary. There has been a good deal of inappropriate modification to cave airflows and spread of misinformation as a result of the 'radon panic' (Ahlstrand and Fry 1978; Yarborough 1976; Hyland and Gunn 1992; Aley 1994).

Cave Carrying Capacity and Alternative Management Concepts

It is probably true for caves that whatever we have today is more than we will have tomorrow. Although new caves are found every year, the growth in leisure time and demand for outdoor recreation outstrip the increase in the resource under pressure. Most caves and their contents are relict features, often formed under climatically more favourable conditions than today. This is especially true of caves in seasonally humid or arid climates, but it also applies to those in more equable climates. Thus to all intents the regeneration of speleothems, for example, may occur on timescales far beyond those of our species.

'The carrying capacity of a cave is zero' (Aley 1976). The concept of carrying capacity was first enunciated in agriculture, for which it was defined in terms of the number of animals that might be placed on a given area on a sustainable basis (Dasmann 1964). The concept was transferred to natural area management by scientists trained in biology, and was widely used for up to two decades (Stankey et al. 1990). The idea rests on the assumption that there is a fixed limit to the use that an area can withstand, and by extension there is also a fixed population which can be supported by that area. This is simply not true. Factors such as the timing and spatial pattern of use, changes in climate and vegetation, and extreme events such as fires and floods result in a changing resource base. Different organisms have differing ecological vulnerability and resilience. The original concept of carrying capacity took no account of socio-economic issues, and did not recognize the non-reversibility of many ecological changes. In particular, many cave animals are habitat specialists and may be vulnerable to minor environmental changes (light, heat, humidity); their populations may not recover from an imposed stress. Thus the concept of carrying capacity is now recognized as a beautiful idea which was murdered by a gang of ruthless facts (Burch 1984).

The issue is now seen as one of identifying the quality of recreational experience which is appropriate to a particular cave environment, and determining the environmental conditions consistent with that use. Thus the question has moved away from just simple numbers of people in caves to the social and environmental conditions which should prevail. Several complementary management tools have emerged. All of them involve the translation of qualitative management goals to quantitative management objectives using environmental indicators and standards.

The first, the Recreation Opportunity Spectrum (ROS), identifies the range of recreation opportunities at a site, the basis for identifying these, and the physical resources needed. This approach requires that caves be classified in terms of their potential uses, likely impacts and resulting environmental conditions. These can range from undisturbed natural environments used for low density recreation (wilderness caves) through to highly modified environments suitable for high density use (tourist caves).

The second, the Limits of Acceptable Change (LAC, Stankey et al. 1985), is concerned with defining those environmental conditions and maintaining them. LAC involves selecting key indicators, setting standards of achievement for each indicator, monitoring to allow comparison to that standard, and modifying recreational use and management strategies in the light of non-conformity to that standard. The first need in the application of LAC is to determine target features or organisms that can be used as indicators. A physical indicator might be the level of breakage or soiling of speleothems in a heavily used passage, or compaction of cave sediments. A biological indicator might be the maintenance of a meta-population of a particular obligate cave species, such as glow-worms or syncarid shrimps. Clearly the choice of an indicator must take account of its representativeness and the feasibility of monitoring it. The need for environmental monitoring implies considerations of replication, frequency (see figure 9.2) and cost. Frequent monitoring of one key parameter is preferable to occasional monitoring of many. It is generally better to monitor an indicator which is simple and cheap to measure at many sites than one so complex and so expensive that it can be afforded only at one or two sites. For example, measuring the evaporation of fixed volumes of water from many open petri dishes in a cave may give a better picture of desiccation problems than a single thermohydrograph near the entrance.

Use of the LAC concept implies a need for visitor monitoring. This may be done simply by visitor logbooks, questionnaires, cheap infra-red beam counters, and fixed photographic reference points. Such monitoring systems must take into account the possibility of vandalism, the intrusiveness of the monitoring apparatus, and, for logbooks, the compliance of visitors (often distressingly low). In many cases a programme of visitor education may repay the effort involved many times over, and may be incorporated into site interpretation.

These two concepts are combined in Visitor Impact Management (VIM). The key elements of the VIM process are:

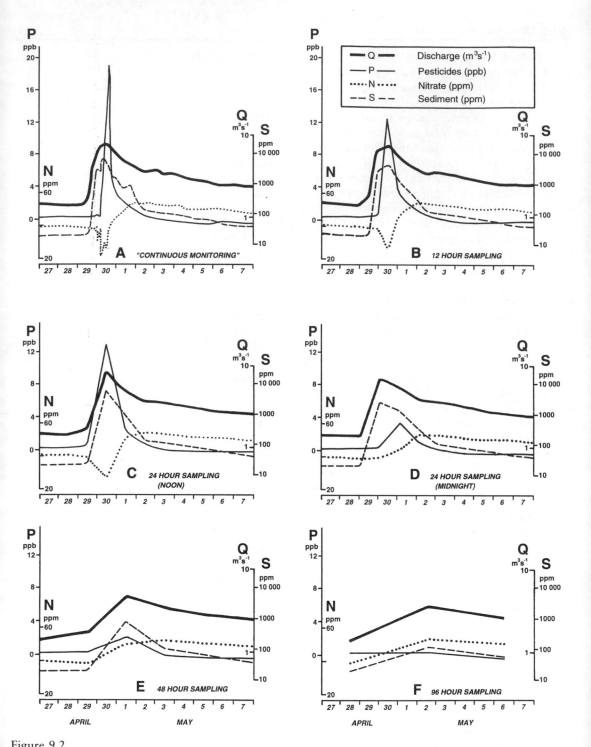

Figure 9.2
Water quality and discharge during a 1984 storm in the Big Spring, Iowa, karst aquifer with conduit flow. Graphs A to F illustrate the effects of differing sampling frequencies, from continuous sampling (real data) up to four-day interval sampling, on interpretation of a flood pulse carrying various pollutants. Note that the pesticides are carried adsorbed onto clay particles and the nitrate is as dissolved load. Redrawn from Quinlan (1990).

- establish precise goals for management and define management units
- identify key indicators which can be used to monitor change
- define desired conditions for these indicators
- design a monitoring programme using these indicators
- establish management responses to monitoring results.

Indicators can be drawn from both the environmental and social sciences. For example, a regular visitor survey could be conducted to determine levels of satisfaction/dissatisfaction with the management of the cave, and the VIM modified accordingly. Populations of keystone cave invertebrates could be monitored seasonally to determine if visitor impacts were causing significant population decline. Or cave carbon dioxide levels could be monitored continuously to see what level was attained after a tour, and how quickly the cave atmosphere returned to normal (the relaxation time). Each site will require a different set of indicators and guidelines, and management must be flexible to take account of this.

Finally, cave managers may have to ask if existing community standards are relevant to caves. In many countries conservation legislation does not specifically cater for caves and karst. It may be necessary to embark on the long and often difficult process of legislative reform. If so, then support from cave user groups (speleologists, youth groups, local tourism boards, scientific societies) may be crucial to eventual success. Prior education programmes for the public may greatly enhance the level of support that is obtained.

Cave Classification and its Applications

From the previous section, cave classification schemes are a vital ingredient in any rational cave management planning. Cave classification needs to consider an individual cave site in relation to the immediate area that surrounds it, to the rest of the park or reserve in which it is located, and to its national and global context. Criteria used to evaluate the significance of an individual cave site (Worboys et al. 1982) include:

- geological considerations – for example, specific features that relate to structure, stratigraphy, palaeontology or mineralogy
- geomorphological considerations, including features that illustrate genetic or chronological relationships, or particularly fine examples of cave morphology
- hydrological considerations, such as the presence of major underground streams or lakes, unusual networks involving breaches of surface divides, or key elements in understanding the conduit network

- biological considerations relating to species richness, the presence of rare and endangered species, unusual trophic structures or key bat maternity sites
- archaeological and cultural considerations, such as the presence of deep, well stratified deposits, the cave's role in the evolution of a regional prehistory, examples of historic cave use such as mining or water management, or its spiritual significance to indigenous people
- geographical considerations of remoteness and wilderness values, proximity to park infrastructure such as roads and camping grounds, recreational opportunities and accessibility from major population centres.

A more detailed discussion of cave and karst significance is provided by Davey (1977).

Once the significance of an individual cave has been assessed (and significance is relative to a particular community viewpoint or management philosophy) then that cave can be placed in one of a number of use categories. Two examples of systems are provided here. One is that used by the National Parks and Wildlife Service of New South Wales, Australia (Worboys et al. 1982):

Group 1 – Closed caves   Those caves in which access is not permitted because of danger from instability or foul air, or caves awaiting classification.

Group 2 – Scientific reference caves   Those caves which are the best representatives of particular attributes of geology, geomorphology, biology or archaeology. The management aim will be to preserve such caves in their natural state so that reference sets of caves and cave life are available in perpetuity. Access will be strictly limited by scientific research permit to experienced cavers. Each permit will indicate the nature and value of the research and its likely impacts on the cave. Party size and experience of users will be specified and impacts should be monitored.

Group 3 – Limited access caves   These caves have such a quality in their physical or biological attributes that they warrant special protection, or have a high level of difficulty which limits cave exploration to very experienced cavers. The principal management aims are to preserve their high quality and/or maintain a high safety standard. Gating is desirable where possible, standards of party size and experience will be strictly maintained, and detailed speleological research is to be encouraged.

Group 4 – Speleological access caves   In this group are caves where the physical or biological attributes do not warrant special protection and the degree of difficulty is suitable for cavers and novices with some training. The principal management aims are to preserve the cave's natural features and provide recreational opportunities for cavers where this does not conflict with its natural values. Gating is usually desirable where possible, party size may be limited and each party must have an experienced or accredited leader.

Group 5 – Adventure caves   These are caves with little or no inherent value other than their morphology, suitable for exploration by inexperienced but properly equipped groups such as organized youth groups, tourist groups led by experienced guides, or novice speleologists.

Group 6 – Public access caves   These include caves in open areas which have been developed as public inspection caves or are suitable for that use, and require no special equipment or clothing unless specified by the ranger. Public access would usually be either with a ranger or as a self-guided tour.

Within large and complex caves further zoning may be necessary, such as that employed by the United States National Parks Service in Mammoth Cave, Kentucky:

Zone A – Closed passages   Closed by gate for scientific or safety reasons.

Zone B – Scientific reference   Part of a cave which may be an excellent example of the geomorphological, geological, biological and/or speleothem attributes of a cave. May be used as a reference against which to evaluate effects of visitor use on the rest of the cave. Usually gated, with access on the basis of a scientific permit.

Zone C – Limited access   Relates to passages or chambers whose physical or biological attributes warrant special protection, or where the degree of difficulty limits exploration to experienced cavers only. Where possible, access is limited by a strategically placed gate.

Zone D – Natural passage   Unimproved passages which may be traversed only by properly equipped and experienced speleologists.

Zone E – Partially developed passages (no electric lighting)   Partially or previously developed passages which are now abandoned. Paths range from good to somewhat primitive. Such passages provide a 'wild' cave experience for untrained visitors with hand-held lights.

Zone F – Fully developed passage (electrically lit)   All passages provided with electric lighting aesthetically arranged and developed with paths, steps, signs, etc. Either self-guided or guided by ranger staff; a fee is charged and the size and frequency of visitor groups is regulated.

Zone G – Intensive use area   Assembly points, visitor facilities and lifts located where possible in areas of the cave with low aesthetic and/or scientific value.

These schemes may serve as the basis for cave classifications elsewhere, bearing in mind that not all zones may be present and that local conditions may render some classes inappropriate. Often such systems of zoning of land use can be interpreted to visitors in such a way that the concepts of rational, ecologically based management are made clear. This is a good way to ensure compliance with such schemes and to increase visitor satisfaction.

Cave Interpretation and its Use in Management

For most of recorded history, caves were associated less with knowledge than with mystery and magic. In the twentieth century the growth of scientific speleology has rendered most caves explicable and valuable components of the natural world. Yet there is still a widespread view that caves are mysterious separate worlds in which romanticism and fantasy have a dominant role. In this context there are some glaringly bad examples of cave interpretation and management. For example, Spotty's Hole at Roker Park, Yorkshire, contains a charming tableau with polystyrene flowstones, mushrooms, elves and a unicorn. Nearby is a small cave with a cardboard elf. At Ranji Cave, Guilin, China, visitors to the cave resurgence are subject to a mock attack by a stone tool-wielding 'caveman' wearing synthetic leopard fur. These extreme examples emphasize the need for subtle and effective interpretation in tourist caves.

As a starting point for any interpretation, the cave must be placed in its context – both environmental and historic. The interconnectedness of the surface and cave environments, the dependence of the limestone solution process on surface water and soil processes, and the ways in which geology influences cavern shape can all be elaborated here. The story of the cave's discovery and its uses over time make interesting material for guides. Greater satisfaction for visitors can be achieved if there is more interaction with the guides. One way of doing this is to let the visitors talk more, ask questions and be asked in return. Often unknown vignettes of local history will be revealed this way. There is nothing worse than an obviously bored cave guide reciting a prepared commentary to a small group and ignoring their questions. There needs to be a coordinated scheme of training in management and interpretation for cave guides and workers, based on a manual which outlines basic speleology at an appropriate level, specifies interesting features of the cave sites being interpreted, and provides management objectives so that the work of the guide can be placed in its context.

An often successful mode of interpretation is to run specific rather than general tours. For example, rather than covering all aspects of the cave, a tour could concentrate on its history, or its geology, or the cave biology. These specialist tours could be run at set times or in concert with more generalized tours. Such a programme is in place at Mammoth Cave, Kentucky, and results in high levels of visitor satisfaction. Adventure tours are a good way of introducing tourists to 'off path' caving where there are suitable caves (e.g., old tourist caves no longer lit, horizontal stream caves, small grottos), adequate staffing and equipment.

Interpretation should ideally focus on concepts rather than on facts. Caves provide us with an unique opportunity to look at the earth and its geology in three dimensions. The rapid transfer of water underground in karst into caves where it is again visible makes it possible for us to examine the effects of land use on groundwater resources in a way not possible in other rock types. The role of vegetation in moderating karst soil and regolith processes can also be incorporated into cave interpretation.

A key element in successful cave interpretation is good, effective lighting (see plate 9.6). Despite a global trend towards the use of natural white light in caves, there are a number of sites which persist in the employment of coloured lights, usually green, blue and red. These detract from the natural beauty and colours of the speleothems. Cave lighting is often the last element of tourist cave infrastructure to be installed, yet it has a profound effect on the visitor's experience of a site. It illuminates that which otherwise

Plate 9.6
Effective lighting in tourist caves is designed to enhance feelings of depth and mystery, and to highlight features of high aesthetic value. Photo of Jersey Cave, Yarrangobilly, New South Wales, by Andy Spate.

would not be seen, and creates the images that visitors take away as memories or photographs. Historical perceptions of caves were associated less with knowledge and understanding than with mystery and marvels. Despite the emergence of cave science as a vibrant discipline in the late nineteenth and twentieth centuries, there is still a widespread community view that caves are separate worlds, disconnected from our sunlit world, in which fanciful stories and lighting are acceptable. This viewpoint was maintained and enhanced by the early practice of using hand-held lights (candles, carbide lamps and torches) for tourist caves. Under these conditions a very small portion of the cave was seen by the visitor, and the experience had qualities of adventure and discovery which provoked long-lasting memories which might be turned into prose or poetry. Technological advances in the first few decades of this century led to the installation of complex lighting systems which illuminated much of the cave along a tourist path, removing the mystery but having a much greater impact on the individual cave on account of the physical infrastructure (wiring, switchboxes, light fittings) and its consequences (lampenflora, heat, disturbance to cave fauna, concrete and wiring leachates, excavation debris). The great financial and labour cost of lighting installation in a cave resulted in these becoming very fixed structures, growing in an ad hoc manner and often overlain on existing structures. Cave managers would have to produce very strong arguments to change an existing lighting and path system in a cave.

In many tourist caves the growth of lighting systems has led to a situation where the cave is almost entirely lit, requiring a minimal lighting control involvement by the guide. In this case both interpretation and educational satisfaction are diminished because there is little interaction with the environment and the control of lighting both for effect and for 'atmosphere' is not possible. Essentially the cave becomes a self-guiding exercise which can cope with large numbers of visitors, regardless of the desirability of doing so. Alternatives to this involve removing lights from caves, installing low-intensity track lights and 'feature' lights which a guide can use selectively for effect and in interpretation. Examples of this approach include the highlighting of a particular formation of great beauty or scientific interest, or using lights to create the impression of passage length or depth. The combination of creative lighting with running water in a cave can be very effective indeed. Flexibility is enhanced, allowing individual guides to tailor the lighting to the theme of a tour. In an interesting return to historic methods, many tourist caves now offer periodic adventure tours where hand-held lights (including candles) or helmet-mounted lights are used by the visitors.

The physical structure of cave paths and signs may have a very great effect on the visitor's experience. Today it is possible to use backlit signs with photographic or graphic material incorporated. This reduces the need for extra lighting on signs.

Many caves discovered in the nineteenth century had narrow passages and delicate formations. There was a major visitor impact very early in the history of their use:

The formations on walls are extremely delicate, some of it white and some like yellow coral. The roof has been slightly defaced by certain nineteenth century cads. In various places the 'mark of the beast' in lampblack has been produced by holding candles near to the ceiling and moving them about gradually. The sooty Hieroglyphics remain to this day, as an evidence of vanity and folly. The floor which was once like alabaster, is now soiled by the tramping of feet. (Cook 1889)

Unfortunately the response of management was to encase the visitor in wire mesh (see plate 9.7) so that contact with formations

Plate 9.7
Many tourist caves developed in the nineteenth century had steel and chicken wire structures to keep visitors away from formations. A number of tourist caves have now removed these unnecessary items, but some remain (Castle Cave, Yarrangobilly, New South Wales). Photo by Andy Spate.

was impossible. Visibility was, of course, greatly impaired as well. Over time this netting and the poles that supported it weathered to produce salts of zinc and iron which stained speleothems, contaminated pools and made paths slippery. The wire netting collected lint from visitors' clothing, and many received small cuts from snagging the wire (Holland 1992). The removal of this wire and associated structures is difficult, with residues inevitably remaining. Old electrical wiring which has deteriorated in the cave poses similar problems of removal, and likewise old wooden structures deteriorate to a fibrous pulp, which provides an extra energy source which may disrupt the cave ecology. Today alternatives for visitor control include carefully placed lighting to create dark zones between decorations and visitors, or the use of infra-red beams which when broken by straying visitors may activate a siren or bell. Carbon glass screens, wide plastic mesh or perspex are viable alternatives for very sensitive areas.

In summary:

- Cave lighting and pathways must be designed to minimize the effects on cave microclimates and biota.
- Subtlety and effective interpretation are the key tools of cave managers.
- Only through public education can the aims of ecologically based cave management be achieved.

Management of the Glow-Worm Cave, Waitomo, New Zealand

The King Country on the west coast of the North Island of New Zealand contains extensive Oligocene limestones in which extensive river caves have formed (Williams 1982), including the well-known Waitomo Caves system. Above the caves the landscape is comprised of hilly pastureland, native forests and small areas of pine plantations. Around 75 per cent of the native forest has been cleared (Williams 1975). There is a mixture of fluviokarst and polygonal karst, with rapid transfer of water underground into large conduits. Thus significant amounts of sediment have made their way into the caves since European settlement in the late nineteenth century. In particular, the water level in that part of the cave used for boat tours has risen over a metre on account of sediment buildup, creating difficulties for navigation.

The ecosystem of the Glow-Worm Cave at Waitomo is a good example of a site where both agriculture and tourism development have had complex impacts and interactions with a karst hydrologic system (see figure 9.3). The lowest level of the diagram represents the external factors impinging on the cave; a change in any of these components will affect the features above it. Careful consideration must be given to the changes resulting from any land use modifica-

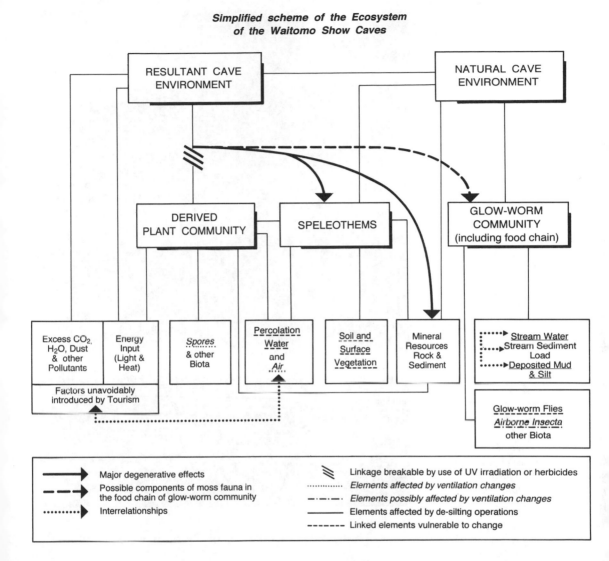

Figure 9.3
Simplified scheme of the
ecosystem of the Waitomo
show caves, showing
factors affecting the cave
ecosystem and critical
linkages for management.
Redrawn from Williams
(1975).

tion to avoid decline in cave ecosystem quality or function. In
particular, the impacts of declining water quality and increased
stream sedimentation were visible on the populations of the glow-
worm Arachnocampa luminosa, whose displays (see plate 9.8)
above a subterranean stream are the major attraction for roughly
400 000 visitors each year. Other impacts of tourism were increased
temperatures and carbon dioxide levels in the drier parts of the
cave, which were causing some re-solution of the speleothems.
Following widespread concerns over the visible deterioration of the
cave up to 1975, a major conservation study was initiated
(Kermode 1974; Williams 1975) and some of its recommendations
have since been implemented (Williams, 1985).

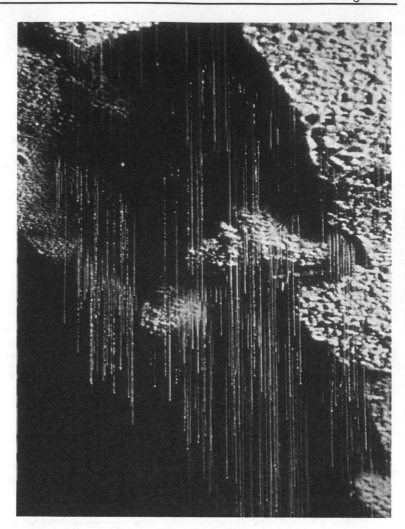

Plate 9.8
Massed threads of
glowworms cover the wall
of the Glow-Worm
Grotto, Waitomo, New
Zealand.

The specific recommendations of the study were as follows:

Cave restoration
a) Reduce excessive sediment supply to the Waitomo
   Stream by immediate planting of disturbed land along
   roads.
b) Reduce the effectiveness of the sediment trap within the
   Glow-Worm Cave by totally removing the artificial rock
   barrier at the resurgence of the cave.
c) Minimize the tendency of the Waitomo Stream to deposit
   sediment within the Glow-Worm Cave by increasing
   stream velocity in the cave as a consequence of reducing
   friction to flow and increasing channel gradient downstream of the cave.
d) Restore a navigable depth within the Glow-Worm Grotto

by pumping out sediment, preferably from the downstream side.

e) Clean hard formations with variable pressure steam jets.

f) Cautiously restore natural ventilation in the caves, but ensure that relative humidity levels stabilize near 100 per cent.

Cave maintenance

a) Reduce sediment supply to the Waitomo Stream by encouraging bush regeneration and reafforestation in the basin of the stream upstream of the Glow-Worm Cave.

b) Trap as much sediment as possible upstream of the Glow-Worm Cave in a sedimentation pond produced by damming the Waitomo Stream.

c) Maintain a navigable water depth within the Glow-Worm Grotto during tourist hours by judicious use of a water control structure at the cave resurgence. Open the structure at all other times to permit free passage of sediment.

d) Maintain clean formations by *absolute minimal* use of lights. Position lights so that they do not shine directly onto nearby formations, except for very short periods.

e) Experiment with lighting in order to find that form of least use to photosynthesizing plants.

Cave management

a) Appoint an appropriately qualified officer within the Tourist Hotel Corporation with specific full-time responsibility for the cave.

b) Guides to be better trained, with professional standards and career structure comparable with the ranger service of the National Parks.

Research

In order to learn how to control cave processes, research should be encouraged, especially on:

a) hydrological and sedimentological processes and rates

b) the ecosystem of the glow-worm

c) artificial cave plant communities and their biochemical effects

d) micro-meteorological and geochemical processes, especially as they affect speleothems (see plate 9.9).

e) lighting systems within tourist caves.

In early 1979 the glow-worm display turned off for three months and resulted in a dramatic fall in visitor numbers. This, and the continuing visible deterioration of the cave, prompted increased activity. A new tourist exit was opened, allowing the closure of cave chambers with excessively high carbon dioxide levels, while a maximum visitor rate of 200 people per hour in the cave was instituted.

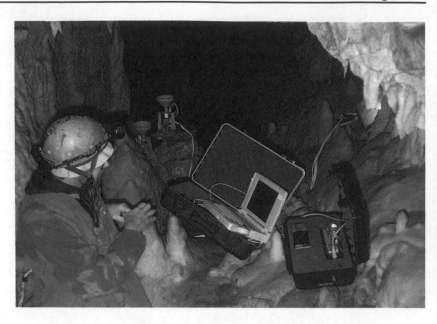

Plate 9.9
Monitoring stalactite drip rates and chemistry in Ruakuri Cave, Waitomo, New Zealand. Tipping bucket gauges with collection funnels are continuously feeding data to Campbell dataloggers, which can be interrogated and downloaded to a computer at regular intervals.

The Waikato Valley Authority set up a comprehensive land capability study in the Waitomo catchment, aimed at maintaining and enhancing water quality and reducing soil erosion. Factors contributing to erosion such as road works, on-farm erosion and feral goats were identified. Controls on further land clearance and on road construction methods were implemented.

The Waitomo area now has a visitor centre and cave museum which is of world class, and tourist cave atmospheres are continuously monitored using dataloggers. Recent developments in adventure tourism at Waitomo have included organized wild caving in the form of 'black water rafting' (McPherson, 1995). Small groups of visitors are trained and equipped and taken on fully guided tours involving cave rafting, canyon traversing and abseiling. These activities are carefully monitored for impacts and the guides are highly trained. The recent return of land ownership to the descendants of the original Maori owners has been a historic step which will add a new cultural dimension to cave interpretation.

References

Ahlstrand, G. M. and Fry, P. L. 1978: Alpha radiation project at Carlsbad Caverns: two years and still counting. In R. Zuber, J. Chester and S. Gilbert (ed.) *Proceedings of the First Cave Management Symposium.* Albuquerque: Adobe Press, 133–7.

Aley, T. 1976: Caves, cows and carrying capacity. *National Cave Management Symposium Proceedings 1975.* Albuquerque: Speleobooks, 70–1.

Aley, T. 1994: Some thoughts on environmental management as related to cave use. *ACKMA Journal* 17, 4–10.

Andrieux, C. 1983: Etude des circulations d'air dans la Grotte de Niaux: consequences. *Karstologia* 1, 19–24.

Bellwood, P. 1978: *Man's conquest of the Pacific: the Prehistory of Southeast Asia and Oceania.* New York: Oxford University Press.

Blackwell, B., Schwarcz, H. P. and Debenath, A. 1983: Absolute dating of hominids and paleolithic artefacts of the cave of La Chaise-de-Vouthon (Charente), France. *J. Archaeo. Sci.* 10, 493–513.

Bonwick, J. and Ellis, R. 1985: New caves for old: cleaning, restoration and redevelopment of show caves. *Cave Management in Australasia VI,* Waitomo, 134–53.

Brunet, J., Vidal, P. and Vouve, J. 1985: Conservation de l'art rupestre. *UNESCO études et documents sur le patrimoine culturel 7.*

Burch, W. R. 1984: Much ado about nothing: some reflections on the wider and wilder implications of social carrying capacity. *Leisure Sciences* 6, 487–96.

Burney, D. A., Brook, G. A. and Cowart, J. B. 1994: A Holocene pollen record for the Kalahari Desert of Botswana from a U-series dated speleothem. *The Holocene* 493, 225–32.

Campbell, J. A. 1977: *The Upper Palaeolithic in Britain.* Oxford: Oxford University Press.

Carlson, K. R. 1993: The effects of cave visitation on terrestrial cave arthropods. *Proceedings 1991 National Cave Management Symposium,* Bowling Green, Kentucky, 338–45.

Cigna, A. 1989: La capacita ricettiva delle grotte turistiche quale parametro per la salvaguardia dell'ambiente sotterraneo: il caso dell Grotte di Castellana. *Atti. XV Congr. Naz. Speleologia. 10–13 Sept. 1986, Gruppo Puglia Grotte – Amm. Com. Castellana Grotte,* 999–1011.

Cook, S. 1889: *The Jenolan Caves: an Excursion in Australian Wonderland.* London: Eyre & Spottiswoode.

Crawford, R. L. and Senger, C. M. 1988: Human impacts on populations of a cave Dipluran (Campodeidae). *Proc. Washington State Entomol. Soc.* 49, 827–30.

Dasmann, R. F. 1964: *Wildlife Biology.* New York: Wiley.

Davey, A. G. 1977: Evaluation criteria for the cave and karst heritage of Australia. *Helictite* 15, 1–41.

Dragovitch, D. and Grose, J. 1990: Impact of tourists on carbon dioxide levels at Jenolan Caves, Australia: an examination of microclimatic constraints on tourist cave management. *Geoforum* 21 (1), 111–20.

De Freitas, C. R. and Littlejohn, R. N. 1987: Cave climate: assessment of heat and moisture exchange. *Journal of Climatology* 7, 553–69.

De Paepe, D. and Hill, C. A. 1981: Historical geography of United States saltpeter caves. *NSS Bull.* 43, 88–93.

Everson, A. R., Chilman, K. C., White, C. and Foster, D. 1987: Recreational use of seven wild caves in Missouri. In J. M. Wilson (ed.) *NSS 1987 Cave Management Symposium.* Huntsville, AL, 11–21.

Gamble, F. M. 1981: Disturbance of underground wilderness in karst caves. *J. Environ. Stud.* 18, 33–9.

Glasser, N. F. and Barber, G. 1995: Cave conservation plans: the role of English Nature. *Cave and Karst Science* 21, 33–6.

Gunn, J., Fletcher, S. and Prime, D. 1991: Research on radon in British limestone caves and mines, 1970–1990. *Cave Science* 18, 63–6.

Hazlinsky, T. 1985: *International Colloquium on Lamp Flora, 10–13 Oct. 1984.* Budapest: Hungarian Speleological Society, 155 pp.

Hill, C. A. and Forti, P. 1986: *Cave Minerals of the World.* Huntsville, AL: Nat. Speleo. Soc.

Holland, E. 1992: Away with wires! *ACKMA Journal* 10, 22–4.

Hyland, R. and Gunn, J. 1992: Caving risks. *New Scientist* 135, 183.

Jablonsky, P. 1990: Lint is not limited to belly buttons alone. *NSS News* 48 (5), 117–19.

Jablonsky, P. 1992: Implications of lint in caves. *NSS News* 50 (4), 99–100.

James, J. M., Gray, M. and Newhouse, D. 1990: A preliminary study of lead in cave spiders' webs. *Helictite* 28 (2), 37–40.

Kermode, L. 1974: Glowworm Cave, Waitomo, conservation study. *NZ Speleological Bulletin* 5, 329–44.

Kiernan, K. 1989: Karst management issues at the Jenolan tourist resort, NSW, Australia. In D. Gillieson and D. Ingle Smith (ed.) *Resource Management in Limestone Landscapes: International Perspectives* Special Publication no. 2, Department of Geography & Oceanography, University College, Australian Defence Force Academy, 111–31.

Kiernan, K., Jones, R. and Ranson, D. 1983: New evidence from Fraser Cave for glacial age man in south-west Tasmania. *Nature* 301, 28–32.

Lyons, R. G. 1992: Radon hazard in caves: a monitoring and management strategy. *Helictite* 30 (2), 33–40.

McCracken, G. F. 1989: Cave conservation: special problems of bats. *NSS Bull.* 51, 49–51.

McPherson, D. 1995: Tourism as a cultural process: a worked example. *NZ J. Geog.* Oct. 1995, 7–15.

Merenne-Schoumaker, B. 1975: Aspects de l'influence des touristes sur les microclimats de la grotte de Remouchamps. *Ann. Spéléo.* 30 (2), 273–85.

Newbould, R. 1976: Steam cleaning of the Orient Cave, Jenolan. *Cave Management in Australia, Proceedings of the First Australian Conference on Cave Tourism.* Jenolan Caves NSW, 10–13 July 1973. Broadway: Australian Speleological Federation, 87–90.

Quinlan, J. 1990: Special problems of pround-water monitoring in karst terrains. In D. M. Nielsen and A. I. Johnson (ed.) *Ground Water and Vadose Zone Monitoring.* ASTM STP 1053, 275–304.

Rajczy, M. and Buczko, K. 1989: The development of the vegetation in lamplit areas of the cave Szelo-Hegyi-Barlang, Budapest, Hungary. *Proc. 8th Int. Congr. Speleo., Kentucky,* 514–16.

Reaich, N. M. and Kerr, W. J. 1991: Background alpha radioactivity in Cuckoo Cleeves Cavern, Mendips. *Cave Science* 18, 83–4.

Ren, M., Liu, Z., Jin, J., Deng, X., Wang, F., Peng, B., Wang, X. and Wang, Z. 1981: Evolution of limestone caves in relation to the life of early man at Zhoukoudian, Beijing. *Scientia Sinica* 14 (6), 843–51.

Slagmolen, C. and Slagmolen, A. 1989: La maladie verte. *Regards* 6, 25–31.

Solecki, R. 1972: *Shanidar: the Humanity of Neanderthal Man.* London: Allen Lane, 222 pp.

Spate, A. and Moses, C. 1994: Impacts of high pressure cleaning: a case study at Jenolan. *Cave Management in Australasia* X. Carlton South, Victoria: ACKMA, 45–8.

Stankey, G. H., Cole, D. N., Lucas, R. C., Petersen, M. E. and Frissell, S. S. 1985: The limits of acceptable change (LAC) system for wilderness planning. *USDA Forest Service Intermountain Forest and Range Experimental Station General Technical Report* INT-176.

Stankey, G., McCool, S. and Stokes, G. 1990: Managing for appropriate wilderness conditions: the carrying capacity issue. In J. Hendee, G. Stankey and R. Lucas (ed.) *Wilderness Management*, 2nd edn. Golden, CO: North American Press, 215–39.

Stitt, R. 1977: Human impact on caves. *National Cave Management Symposium Proceedings 26–29 October 1976*. Albuquerque: Speleobooks, 36–43.

Tercafs, R. 1988: Optimal management of karst sites with cave fauna protection. *Environmental Conservation* 15, 149–58.

Villar, E., Fernandez, P. L., Gutierrez, I., Quindos, L. S. and Soto, J. 1986: Influence of visitors on carbon dioxide concentrations in Altamira Cave. *Cave Science* 13 (1), 21–3.

White, W. B. 1988: *Geomorphology and Hydrology of Karst Terrains*. New York: Oxford University Press.

Williams, D. R. 1985: The future of cave tourism in Waitomo. In D. Williams and K. A. Wilde (eds) *Cave Management in Australasia* 6, 13–22.

Williams, P. W. (ed.) 1975: Report on the conservation of Waitomo Caves. *NZ Speleological Bulletin* 5 (3), 373–96.

Williams, P. W. 1982: Karst landforms in New Zealand. In J. Soons and M. J. Selby (ed.) *Landforms of New Zealand*. Auckland: Longman Paul, 105–25.

Williams, P. W. 1993: Environmental change and human impact on karst terrains: an introduction. In P. Williams (ed.) *Karst Terrains: Environmental Changes and Human Impact: Catena Supplement* 25, 1–20.

Worboys, G., Davey, A. and Stiff, C. 1982: Report on cave classification. *Cave Management in Australia IV*, 11–18.

Yarborough, K. 1976: Investigation of radiation produced by radon and thoron in natural caves administered by the National Park Service. *Proc. Nat. Cave Manage. Symp.* Arkansas, October 1976, 703–13.

Yuan, D. 1991: *Karst of China*. Beijing: Geological Publishing House, 224 pp.

TEN

Catchment Management in Karst

Introduction

Throughout this book the connectivity of caves and their overlying karst has been stressed. Conservation of caves is a very short-sighted preoccupation unless it is accompanied by energetic attention to conservation of karst as a whole. Nearly all of the karst solution process is moderated by factors operating on the surface of the karst, in the epikarst and in the subcutaneous zone. Surface vegetation regulates the flow of water into the epikarst through interception, the control of litter and roots on soil infiltration, and the biogenic production of carbon dioxide in the root zone. The metabolic uptake of water by plants, especially trees, may regulate the quantity of water available to feed speleothems. Trees in particular are like large carbon dioxide pumps, releasing 20 to 25 per cent of the atmospheric gas uptake through root respiration (Aley 1994). Thus clearfelling of karst, or major changes consequent on plantation establishment, may radically change the flow and quality of water in the karst. Soil erosion in excess of the natural rates may infill streamsinks, dolines or caves, and may totally smother cave life. Changes to surface drainage resulting from contour banking, irrigation or river regulation may interrupt or drastically reduce the supply of karst water. The release of fertilizers, herbicides and insecticides from agricultural activities may compromise cave ecosystems beyond their capacity to recover. Water is the primary mechanism for the transferral of surface actions to become subsurface impacts.

Basic Concepts in Karst Management

Karst management must be holistic in its approach and should aim to maintain the quality and quantity of water and air movement through the subterranean environment as well as the surface. Managers of karst areas should recognize that these landscapes are complex three-dimensional integrated natural systems comprised of rock, water, soil, vegetation and atmosphere elements. In all cases the intention should be to maintain natural flows and cycles of air and water through the landscape in balance with prevailing climatic and biotic regimes. All users of karst terrains should recognize that, in karst, surface actions may be rapidly translated into impacts underground and elsewhere.

Pre-eminent among karst processes is the cascade of carbon dioxide from low levels in the external atmosphere through greatly enhanced levels in the soil atmosphere to reduced levels in cave passages. Elevated soil carbon dioxide levels depend on plant root respiration, microbial activity and a healthy soil invertebrate fauna. This cascade must be maintained for the effective operation of karst solution processes.

Thus the integrity of any karst system is dependent upon a specific relationship between water and land; this water is often drawn from a very wide catchment area, and any alteration in the hydrologic system will threaten the karst and those caves which have a continuing relationship to the water levels. However, many caves, abandoned by the original formative waters as groundwater levels have been lowered, will be relatively dry, relatively static in character and essentially non-renewable. Their contents – formations, sediments and bones – are especially vulnerable, and may need special management provisions.

Climate change has occurred over the geological timescales within which karst systems have evolved. Human intervention has the potential to alter climate in ways that may radically affect natural karst processes. Management prescriptions must be flexible, must recognize this possibility and must maximize the resilience of the system. The effects of high magnitude–low frequency events such as floods, fires and earthquakes must be perceived in management strategies at regional, local and site-specific scales.

Defining Karst Catchments

The catchment of a karst drainage system is usually much larger than just the area of limestone outcrop and the obvious non-karstic contributing catchment. However, defining the contributing catchment of a cave may be difficult if not impossible in some cases. A minimalist approach would be to define the catchment as the area of limestone outcrop. This is often convenient for agencies wishing to restrict the application of environmental protection legislation to a given situation such as a quarry or landfill. This neglects the possibility that the limestone is continuous though not outcropping in a given terrain, or that surrounding non-karstic rocks are contributing significant quantities of water by surface or subsurface flow. In many cases a thick mantle of colluvium lies over the limestone and directly feeds cave systems. This is especially true in areas which were formerly glaciated or which have been subject to repeated mass movements over geologic time. The complexity of karst drainage systems has been illustrated for the Mammoth Cave area in Chapter 2. The elucidation of this drainage network was the result of over twenty years' investigation and hundreds of dye-tracing experiments. It is rare for managers to have this level of detail. Usually, decisions must be made on the basis of a few dye-tracing experiments, aided by local geological knowledge and intuition. Karst managers must therefore learn to expect the unexpected!

Subterranean breaches of surface drainage divides are often the norm rather than the exception, and the exact conditions for the activation of conduits may depend on storm events or antecedent rainfall. In some cases palaeokarst conduits may be reactivated. Thus the definition of a karst catchment is imprecise and must have a dynamic boundary to take account of extreme events. This is best achieved by constructing buffer zones around limestone massifs, in which any change to land use must be preceded by investigations of the drainage network and its dynamics using repeated dye-tracing experiments.

For karst areas, the concept of total catchment management (see plate 10.1) becomes more vital than in many other lithologies. This involves the coordinated management and utilization of physical resources of land, water and vegetation within the boundaries of a catchment to ensure sustainable use and to minimize land degradation. Proper environmental management on karst terrains rests on a base of public acceptance that clear linkages exist between surface and underground systems, and that these linkages are of fundamental importance to karst system function.

Plate 10.1
Karst catchments, such as in this polygonal karst area at Waitomo, New Zealand, will often extend beyond the limestone boundary and are especially vulnerable to altered land use and water pollution. Within this catchment land uses include grazing, forestry and quarrying.

Vegetation and Caves

The thin mantle of soil on most limestone areas has a great significance for karst processes, as we have seen in Chapter 2. First, there are a number of interactions between limestone soils and vegetation. The free vertical drainage of most limestone soils creates special conditions for evapotranspiration, gas exchange and root penetration. Large eucalypts act as water pumps, taking up 250 to 270 L day^{-1} for each tree of river red gum (*Eucalyptus camaldulensis*). Tree roots can penetrate to depths of 30 to 50 m in search of water, especially so in the seasonally humid climates of northern Australia. Litter retention is of great importance for nutrient cycling, with most release by decomposition occurring within six months of litter fall.

Secondly, the vegetation structure (especially projected foliage cover) is important for interception of rainwater, soil water infiltration and temperature control in soil and subcutaneous zones. This directly affects both the quantity and quality of water available as feed water for speleothem growth in underlying caves. The penetration of tree roots, and the release of complex organic acids and phenolic compounds by them, aids the enlargement of bedrock fissures in karst and ensures the high degree of secondary porosity characteristic of limestone terrains.

Native plants growing on limestone are frequently displaced by exotic species, creating management problems. Most plant species can be classified according to their life strategies – their life-forms, resistance to disturbance, rates of seed production and viability. Highly competitive species such as blackberries grow rapidly, resist disturbance, bear copious viable seed and may inhibit the establishment and growth of other species near them by poisoning the soil (allelopathy). Stress-tolerating plants such as figs and eucalypts have adaptations to ensure adequate water supply (deep roots, high root-shoot ratios), to resist fire (resistant bark, lignotubers, and vegetative reproduction) and to exploit low levels of nutrients (root fungi associations, high root hydrogen ion production). Ruderal species, those that habitually colonize disturbed habitats such as roadsides, have tuberous energy stores, very effective seed dispersal, and easily pollinated flowers. Dandelions and St John's Wort are good examples of ruderal plants. In general stress-tolerating plants are more numerous on limestone soils, with some ruderals present, but highly competitive species are rare. In part this explains why many karst areas have been colonized by competitive exotic species which naturally grow in other limestone areas. Those native species which are stress-tolerators are very susceptible to displacement by exotic species when the disturbance regime is altered radically. Their reproductive strategies may be adapted to infrequent disturbance and they may be unable to complete their reproductive cycle if repeatedly stressed. This may happen as a result of increased fire frequency, after earth-moving operations or after accelerated erosion. In the case of fire, the transference of prescriptions derived for one area to another may not produce the same response in vegetation. A change in the vegetation may therefore have major consequences for karst hydrology and the limestone solution process. This has been noted for the replacement of native woodland by exotic pines with higher water use and a tendency to acidify the soil, resulting in interruption of growth and even erosion of speleothems in caves below the introduced forest. In many karst areas old growth native forests have been cleared and replaced with monospecific plantation forests, often coniferous. These plantations have higher basal area and often higher water demand per hectare than the forests they replace (Costin et al. 1984). Thus there may be a reduction in the flow of percolation water to the karst system, as well as some sediment transfer associated with felling and roading. At Naracoorte Caves, South Australia, exotic pine plantations over caves were cleared in 1990; already speleothems deprived of feed water for decades have reactivated and extended. At Yarrangobilly in the Snowy Mountains of Australia, eucalypt forests were cleared in the 1930s for exotic pine plantations. Caves underlying the pine forest had high root biomasses visible and were relatively dry. Since clearance of the pine and partial regeneration of native vegetation, some cave formations have reactivated. Soil loss in the area has been minimal, with most sediment transfer occurring in spring (NPWS 1983).

Accelerated Soil Loss in Karst

There are numerous examples of accelerated soil erosion on karst areas worldwide (Gams et al. 1993; Urushibara-Yoshino 1993; Yuan 1993; Urich 1991; Gillieson 1989; Kiernan 1988). Limestone soils tend to be shallow and stony, with low to moderate nutrient holding capacity because of excessive leaching as a result of free drainage. There is thus a strong tendency for devegetated or heavily used limestone soils to erode down to bedrock surfaces quite rapidly. This soil stripping (see plate 10.2) can be seen in the classic karst of the Burren, Ireland; the Dinaric karst; the Guizhou polygonal karst of China; and the karst of Vancouver Island, British Columbia. The process started some 2000 years BP in Greece and continues today in areas like the Bohol cone karst, Philippines. Eroded soil material is rapidly transferred underground to block passages, divert or impound cave streams, or smother cave life (Kranjc 1979). Soil erosion control is therefore a high priority for karst managers, and much depends on the effectiveness of revegetation. There is usually accelerated soil loss and tree decline associated with forestry operations on karst. On Vancouver Island, forests clearfelled since 1900 have regained only 17 per cent of the original timber volume after 75 years, and soil depth loss ranges

Plate 10.2
Stripped limestone pavement at Malham Cove, Yorkshire, shows the legacy of agricultural land use from Neolithic times onwards.

from a mean of 25 per cent five years after logging to 60 per cent after ten years. Clear guidelines for forestry operations on karst need to be developed and adhered to. Soil management in karst must aim to minimize crosive loss (see plate 10.3) and alteration of soil properties such as aeration, aggregate stability, organic matter content and a healthy soil biota. Pivotal to the prevention of erosion and maintenance of critical soil properties is the presence of a stable vegetation cover.

Erosion prediction and control on limestone soils follows the same provisions as most other soil types with some important exceptions. Water erosion by sheet and rill processes may be predicted at a point using the Universal Soil Loss Equation (USLE). This has been calibrated for some soils from long-term erosion plot studies (Rosewell and Edwards 1988). Annual soil loss in tonnes/ hectare (A) is a function of the rainfall intensity or erosivity (R), the soil erodibility (K), the slope angle and length (L, S), vegetation cover (C) and cultivation practices (P):

$$A = f(R \times K \times L \times S \times C \times P) \qquad (10.1)$$

Since soil erosion is a function of ground slope and slope length, any factors locally increasing the slope angle will result in more erosion. On non-karstic terrain, the slope angle and length are functions of the slope distance to the nearest drainage line. On limestone the intimate connections between the surface and the underground cave conduit network lead to locally steeper hydraulic gradients which enhance erosive processes. There is also greater potential for vertical abstraction of material down joints and fissures by sinkhole collapse, gullying, or soil stripping (see plate

Plate 10.3
Steep (30 to 40°) gardens cultivated for sweet potato in the Baliem Valley, Irian Jaya. These limestone slopes are experiencing very high rates of soil erosion under the combination of intensive use, high rainfall and friable soil.

Plate 10.4
Once stripped of soil, complex subsurface limestone solution forms are exposed, allowing the landowner's artistic talents free expression. Near Parker Cave, Kentucky.

10.4). The preservation of adequate vegetation cover and of soil structure (reducing erodibility) assumes greater importance on limestone than elsewhere in mitigating this effect.

Caesium-137 has been used extensively as a tool to measure rates and patterns of soil redistribution. Fallout of this radionuclide is derived from atmospheric testing of atomic weapons during and after the 1950s, with the 1958 concentrations of ^{137}Cs in soil being still detectable today within Australia. After reaching the soil surface ^{137}Cs has been shown to be irreversibly bound to soil particles, especially clays. Its fate and redistribution within the landscape thereafter occurs only by soil transport processes (Campbell et al. 1986). Its properties as a unique and discrete label of surface soil make it ideal as a tracer to test comparative rates of erosion, in which measurements of areal concentration of ^{137}Cs in erosion or deposition sites are compared with the known total or 'reference' fallout at that location (Murray et al. 1990). Increases in this isotope compared with the 'reference' value indicate areas of soil accretion, and areas of decrease represent zones of erosion. These proportional differences may be converted to quantitative loss by a variety of models (Walling and Quine 1990; Gillieson et al. 1994). The measurements allow us to make estimates of erosion integrated over a long time and thus free of the biases of short-term sampling, which may miss major events in the landscape.

The literature of soil science is extensive and often complex. Under the widely used USDA Soil Taxonomy a very large number (>70) of soil properties must be measured before a soil can be classified and its behaviour predicted. This is also true for soil classifications based on soil formation or genesis, such as Great Soil Groups. A simpler factual key for the classification of soils (Northcote 1979) may be used in which a few key properties are measured in the field. These are the soil profile form, the texture of the visible layers, their pH and colour. From these much of the behaviour of soils can be inferred:

- the profile form tells us about soil drainage, rooting ability of trees and aeration
- soil texture tells us about the water holding ability and nutrient retention
- pH tells us about nutrient availability and the acidity of soil water
- colour tells us about organic matter content, clay minerals and precipitates.

To these should be added total organic matter content, soil water infiltration capacity and penetrometer resistance, a measure of compaction. These are all useful attributes of limestone soils which can be measured in the field and provide a good basis from which to predict soil behaviour, especially susceptibility to erosion and suitability for plant growth. They should be part of the basic toolkit for karst managers.

Agricultural Impacts

Pollutants readily enter karst drainage systems and are rapidly transmitted in cave conduits. The range of likely pollutants and their relative importance are given in table 10.1. In Britain some 148 licensed landfill sites are located on limestone (Chapman 1993, 197). Many of these take industrial wastes, usually because of their remote location. Leachates from these sites may travel to contaminate underground watercourses and springs over several kilometres (see plate 10.5). In Ireland up to 100 000 m³ of slurries from feedlot operations are discharged into limestone aquifers via sinkholes, quarries and caves. This nutrient boost to underground waters may cause dramatic changes in cave ecosystems and severely affect water quality for people. In Slovenia typhoid-paratyphoid epidemics as a result of sewage pollution of limestone groundwater affected populations in the classical karst between 1925 and 1975. Chapman (1993, 198–9) provides further graphic examples of sewage pollution in British karst areas.

Table 10.1   Sources of water pollution in caves (modified from White 1988, 389)

Source	Oxygen demand	Nitrogen, phosphates	Chlorides	Heavy metals	Hydrocarbons, organic complexes	Bacteria, viruses
Domestic and municipal wastes						
Septic tanks	•••	••				•••
Outhouses or privies	•••	••				•••
Sewage lines	••	•				••
Landfills	•••	•	•	•••	••	••
Dumps in sinkholes	•	••	•	••	••	•••
Agricultural wastes						
Feedlots	•••	•••				•••
Fertilizer leaching		•••				
Insecticides and herbicides					•••	
Construction and mining						
Salting of roads			•••			
Mine tailings				•••		
Car park runoff			•••	•	••	
Oil fields			•••		•••	
Industrial						
Petroleum storage					•••	
Chemical dumps		•		•••	•••	
Chemical wastes			•	•••	•••	

Note: number of dots indicates very approximate severity of pollution threat.

Plate 10.5
Turbid water derived from hillslope erosion entering the clear karst spring water of Wee Jasper Creek, New South Wales. Photo by Andy Spate.

Documentation of these probably commenced with the 1845 cholera epidemic but continue to the present day. The comparatively rapid transmission of groundwater flows in karst provides little opportunity for natural filtering or other purifying effects, and so problems such as disease transmission may arise much more readily than in other terrain. Again, the source of the pollution may be located far outside of the karst area itself but can still have devastating impacts. Management should aim to maintain the natural transfer rates of fluids, including gases, through the integrated network of cracks, fissures and caves in the karst. The nature of materials introduced must be carefully considered to avoid adverse impacts on air and water quality.

Limestone and marble are quarried worldwide and used for cement manufacture, as high grade building stone, for agricultural lime, and for abrasives. Most resource conflict over limestone mining revolves around visual and water pollution, as well as loss of recreational amenity and conservation values. In Britain some important caves have been destroyed by quarrying, and this industry remains one of the most potent threats to caves in that country. In the Guilin tower karst of southern China there are numerous small quarries operated for cement manufacture and for industrial fluxes. These, coupled with devegetation as a result of acid rain from coal burning, have scarred many of the towers around the city. Ironically much of the cement is used to expand the infrastructure of tourism – hotels and shops – for people who have come to see the famous karst landscape.

Limestone bodies with high relief are ideal for mining (see

Plate 10.6
The limestone quarry at Mount Etna, central Queensland, has operated for twenty-five years and in that time has destroyed or affected many well decorated caves. One such cave, now protected, is home to half a million insectivorous bats. The mountain is now reserved as a karst national park.

plate 10.6) and are often the most cavernous, and there is often conflict and compromise when there is a high community expectation of continued access to this resource as well as a strong conservation movement. Recognizing that the extraction of rocks, soil, vegetation and water will clearly interrupt the processes that produce and maintain karst, such uses must be carefully planned and executed to minimize environmental impact. Extractive industries may be incompatible with the preservation of natural and cultural heritage.

Meiman (1991) has related changes in land use practices in Kentucky to water quality decline in Mammoth Cave National Park. Flood pulses from heavily used agricultural land produced significant rises in turbidity, chloride, faecal coliform levels and triazine herbicides. Hardwick and Gunn (1993) have summarized land use impacts on British caves (see table 10.2). Of the 79 caves identified, 68 per cent were affected by agricultural pollution: 35 sites were affected by rubbish tipping, 16 by farm tipping, and 14 by infilling of caves.

Dolines and open shafts in karst have long been seen as 'good places' to dispose of rubbish. The 'out of sight, out of mind' principle applies here (see plate 10.7). Such places have commonly

Table 10.2 Recorded agricultural and non-agricultural impacts on British caves (after Hardwick and Gunn 1993)

Area	Agriculture		Number of caves	Non-agriculture
	Number of caves	Impact type		Impact type
Northern Pennines	3	Entrance infilled	4	Fly ash tipping
	4	Closed (water supply)		
	7	Farm tipping		
	3	Carcass disposal		
	4	Scrap metal/car bodies		
	1	Oil waste		
	7	Land drainage		
	1	Farm sewage		
Total	30		4	
Peak District	4	Farm tipping	2	Infilled
	1	Farm sewage	7	Fly ash tipping
			1	Oil (quarrying)
Total	5		10	
Wales	1	Entrance infilled	6	Fly ash tipping
	1	Farm tipping	1	Sewage
	1	Used as byre	2	Oil (quarrying)
Total	3		9	
Scotland	1	Farm tipping	1	Fly ash tipping
Total	1		1	
Mendips	8	Entrance infilled	2	Fly ash tipping
	11	Farm tipping	1	Oil
	6		1	Industrial effluent
Total	25		4	
Devon	0		2	Fly ash tipping
			2	Official tipping
			1	Chemical dumping
Total	0		5	

been used for the dumping of animal carcasses, and this poses a considerable health risk. At Earls Cave, Mount Gambier, South Australia, more than 5000 sheep carcasses were dumped in the cave entrance. This connects directly to the karst groundwater which is used for town water supply (see plate 10.8). Blood and unwanted body parts from slaughterhouses in Kaua'i Island, Hawaii, have in the past been disposed of in caves, altering the cave ecology in favour of exotic carrion feeders and scavengers. In the Chimbu region of New Guinea, the vanquished in a tribal fight were com-

Plate 10.7
Domestic rubbish is often dumped in cave entrances, and provides both an energy source and potential pollutants for cave organisms. Dip Cave, Wee Jasper, New South Wales. Photo by Stefan Eberhard.

pelled to suicide by jumping down the 60 m shaft of Nombe at arrow point. The water from this site drains to major springs used for villages near the town of Chuave.

Fire Management in Karst

One of the more vexed questions about karst catchment management is concerned with revegetation and fire control. Fire management on limestone areas is a contentious subject, especially when severe wildfires have caused loss of life or property. In traditional societies fire is widely used as a vegetation clearance tool (Gillieson et al. 1986; Urich 1991; Head 1989; O'Neill et al. 1993). Most karsts have a low natural fire frequency on account of the shielding effects of limestone outcrops, reduced ground cover and often a more mesic canopy with rainforest elements in the flora. In the impounded karsts of Eastern Australia natural fire frequencies are poorly documented, but the fire interval may be 35 to 50 years or greater (Williams et al. 1994; Holland 1994). Under these conditions relict vegetation types may survive – for example, the vine thickets of North Queensland or the monospecific *Acacia* scrubs on the Bendethera karst, New South Wales (Davies and Gillieson 1986). In Britain there are rare fern species which are now restricted largely to limestone outcrops (Goldie 1993). In these karsts sediment transport occurs only immediately after fires, with minimal soil erosion in the intervening periods.

Hazard reduction burning is widely used by land managers, but may have major deleterious effects on karst areas. In Australia

Plate 10.8
The cenote at Earls Cave,
Mount Gambier, South
Australia, was the site of
dumping for more than
5000 sheep carcases.
Casual dumping still
occurs, polluting the karst
aquifer which serves many
farms and towns.

many authorities aim to burn individual areas on a five- to seven-year cycle. This increased frequency reduces the fuel load but promotes more fire-tolerant vegetation, principally shrubs which are often more flammable than the understorey of grasses and herbs that is replaced. Thus there is potential for changes to the hydrology of the epikarst. Although there may be a management prescription to avoid burning limestone outcrops, there may be inadvertent escape of fires into sensitive areas on account of weather changes. There is a possibility of increased cave sedimentation (Stanton et al.

1992). A careful zoning of fire management, aided by precise mapping of past fire boundaries with buffers around karst areas, may help to reduce these impacts. Elucidation of fire histories using sedimentary charcoal in caves is a promising avenue for research (Holland 1994).

Guidelines for Karst Management

Many land management authorities now have guidelines for operations in karst areas under their jurisdiction. A good example is provided by the Tasmanian Forestry Practices Code (Forestry Commission 1987), in which principles for forestry activities in karst areas are outlined (see table 10.3).

Similarly, the US Forest Service has a directive for management of caves in forestry areas (Forest Service 1986, Directive 2356) which outlines the Acts under which cave management takes place, defines a clear policy for the conservation of caves, and has management prescriptions. These include:

- vegetation manipulation around cave entrances and within the cave's hydrological setting
- activities which might alter entrances or affect cave climates
- introduction of pesticides, herbicides, fertilizers and other deleterious materials either directly into the cave or within the cave's hydrological setting
- changes in water quality and quantity, including pollution by septic systems and landfills
- construction of surface facilities such as roads, buildings, pipelines and parking areas
- extraction of minerals
- activities affecting the food chain and critical habitat of cave life
- runoff water from roads and parking areas
- blocking or changing natural water percolation as a result of compaction, paving or vegetative management
- activities affecting the air quality of a cave.

The directive includes cave development plan provisions, public education, monitoring, and cooperation with other agencies and scientists. The management framework is there, is flexible and recognizes caves as assets rather than liabilities. There is a similar cave resources management prescription for areas under the control of the Bureau of Land Management (BLM 1987), which outlines objectives, policies, and individual responsibilities. The implementation of such directives may, of course, vary considerably in reality depending on staffing, logistics and funding. But there is a clear institutional recognition of the karst system catchment (or hydrologic setting) and its physical and biotic interactions.

Table 10.3 Basic approach for forestry activities in karst areas of Tasmania

Activity	Soil type/erosion class	
	Thin residual karst soils, sand mantles: high erosion class	Thick consolidated soils, compacted clay rich mantles: average erosion class
Earthmoving operations and roading	• Restricted to dry season and suspended in wet weather conditions. Special attention to drainage is required.	• Restricted only during periods of heavy rain and for necessary periods thereafter.
Logging	• All logging in karst catchments will be planned to take account of karst values. • Dry season logging only; suspended in wet weather. • Clear-cut coupes to be kept small with a short fall line dimension. • Cable systems hauling uphill for slopes 15–35%. No logging on slopes >35%. • Snig tracks will be planned to reduce soil disturbance. Snig tracks should not cross mapped caves near the surface, enter any karst depression or divert a natural watercourse. • All known sinking streams, intermittent or ephemeral surface channels, caves or sinkholes will be avoided during logging operations.	• Normal wet weather limitation will apply • Clear-cut coupes will not exceed 200 hectares. • Conventional ground skidding and cable systems. Cable systems above 35%. • Snig tracks should be fully planned and integrated with road plans to reduce soil disturbance. • Mapped caves or sinkholes will be avoided during logging operations.
Post-logging restoration	• Apply cross-drain standards for high erosion class throughout the logging area. Sediment traps may be required.	• Adjacent to sensitive areas apply high erosion class drainage standards.
Loading	• Loading should be done by boom-type loaders. • Landings are to be of minimal area and well removed from karst reserves and depressions. • Landings will be drained into sediment traps which are properly maintained.	• Landings should be properly drained at all times.
Silviculture	• Slash burning on slopes >20% will be avoided.	• No restriction on burning logging slash.
Pollution	• Waste materials will not be dumped in karst depressions or sinkholes. • Poisoning of flora and fauna will no be done adjacent to karst streams or known cave entrances.	
Plans	• Timber harvesting plans will specify karst features for an area and the protection measures to be adopted. Liaison with speleologists should be considered in plan preparation.	

Major differences in karst management may occur between public and private lands. Legislation applicable to private lands may be less prescriptive and there may be a local reluctance to enforce relevant legislation where breaches occur. In some areas community groups concerned with cave use or conservation may exist and should be involved in either the consultative process or the development of cave management plans. Today this occurs in many countries through national speleological federations (Davey 1977; Glasser and Barber 1995). Individuals have an important role to play in cave management through lobbying, education, and direct conservation action where appropriate.

Management is therefore less straightforward when caves occur on private land or where protected areas (such as national parks) include urban areas or diverse land uses. Under the British Wildlife and Countryside Act (1981), Sites of Special Scientific Interest (SSSI) may be designated; there are at present 48 cave SSSIs. The boundaries of the cave SSSIs include over 30 per cent of cave entrances and about 70 per cent of the 800 km of mapped passages in the UK. Waltham (1983) provides a list of these sites, while Chapman (1993, 207) discusses the process in more detail. This notification or listing becomes a charge on the site such that regional authorities are required to inform owners and occupiers of the nature of the scientific interest and the defined potentially damaging operations (PDOs) which might impact on the site. The PDOs are supposed to control operations which fall outside local planning provisions. Currently there are twelve PDOs affecting caves:

- dumping, spreading or discharge of a material
- afforestation or deforestation
- changes in land drainage, including field under-drainage or moor-gripping
- modifications to the structure of water courses
- water extraction, irrigation or other hydrological changes
- extraction of minerals
- construction, removal or destruction of roads, earthworks, walls and ditches, laying or removal of pipes, cables, etc.
- storage of materials in potholes, caves or their entrances
- erection of building structures or engineering works
- modification of natural or man-made features
- removal of geological specimens
- recreational activities likely to damage features of interest.

Any change in land use that might involve a PDO must then be negotiated; as a last recourse a banning order on the proposed activity (a Nature Conservation Order) might be applied. Nearly all the cave SSSIs have been declared on the basis of geology or hydrogeology; clearly sites of biological interest need to have greater representation than at present, although these may be quite small caves. A recent development (Glasser and Barber 1995) has been the enrolling of individual speleological groups in the drawing up of plans of management for designated caves. This type of user and community involvement at a grass-roots level is to be applauded.

Conservation Issues in Karst

Karst waters can be viewed as types of wild rivers where the drainage network is not as obvious as in surface streams, and there is complexity in hydrological linkages and in flow regimes. In many mountain areas the highest parts of karst catchments are still forested and inaccessible. In such areas both water quantity and quality are maintained along with the integrity of ecosystems. It is estimated that one-quarter of the world's population gain their water supplies from karst, either from discrete springs or from karst groundwater. The maintenance of water quality in karst can be viewed as a common good which is becoming increasingly important in those areas where rural populations are increasing rapidly and settlement of karst is well established. In other areas, such as China and the Philippines, recent settlement of karst terrain is creating both opportunities and constraints for sustainable management of karst resources, especially water and soil. Establishment and maintenance of karst protected areas can contribute to the protection of both the quality and quantity of groundwater resources for human use. Catchment protection is necessary both on the karst and on contributing non-karst areas. Activities within caves may have detrimental effects on regional groundwater quality.

The rugged nature and physical isolation of most karst landscapes ensure that they act as refugia for rare and endangered species, such as leaf-eating monkeys in Vietnam and cave-adapted animals such as salamanders and blind fish in North America. Karst landscapes have a high importance for the maintenance of *biodiversity*, and there are still opportunities to protect the habitats of these organisms. In many ways karsts are buffered against climate change, and their biota are less vulnerable in climates characterized by high natural variability. There is often a high degree of endemism in karst biota. Many karst areas preserve relict populations of organisms whose distributions were much greater during past colder or wetter climatic regimes.

Because of the importance of karst areas as biological refugia, further fragmentation by roading and similar activities should be avoided. If they are unavoidable, then corridors for animal dispersal should be maintained as a high priority. Within caves both terrestrial and aquatic fauna are best protected by the preservation of air and water quality. Accelerated stream siltation and compaction of sediments by visitors may be detrimental to cave fauna. The infrastructure of tourist caves (paths, steps, lights) should be designed to avoid decomposition and the release of either toxic substances or additional energy sources into the cave environment.

Caves have long been used for shelter, for religious purposes, for hunting, and as objects of veneration and awe. For indigenous people living with traditional lifeways, access to cave resources is but one component of a complex of land use and rights which is central to their being. For the Tifalmin people of the Star Mountains, New Guinea, caves are the resting places of their ancestral female deity Afekan, and pits in Selminum Tem Cave are of great significance (Gillieson 1980). Cave entrances and streamsinks also mark the boundaries of hunting territories. On the Nullarbor Plain, the flint mines of Koonalda Cave have been used by more than 600 generations of Aborigines, while other caves mark dreaming sites on a mythologically linked songline from the West Kimberleys to Eucla on the Great Australian Bight. In many areas traditional owners are now in partnership with government agencies concerned with the management and interpretation of caves; for example, the Berawan and Penan people of Gunong Mulu National Park, Sarawak, are employed as cave guides, boatmen and in daily site management (Gill 1993). Aboriginal cooperatives in the West Kimberley karst of Australia are running ecotourism ventures involving cave visits, while there are several leaseback arrangements where National Park agencies lease the land from traditional owners who sit on the board and are employed as rangers (Young 1992). Their involvement adds a whole new dimension to interpretive programmes. This form of community involvement in cave management deserves to be more widely employed, even in non-traditional societies.

Rehabilitation and Restoration of Caves

In general, karst systems develop over geological timescales which must inevitably include significantly different environments to that of today. Some karst systems will thus be so out of phase with prevailing conditions that they have no capacity to regenerate. Other systems may have some capacity to regenerate but this may entail timescales greater than that of individual human generations. Caves and their contents (formations, sediments and bones) may have been formed or emplaced under different climate regimes and may remain unaltered for millennia. These may require specific management attention because of their fragility. However, the political process which drives cave rehabilitation is operating on a short timescale, often the three to five years until the next election. Management agencies must produce results for politicians who are responsible to community groups and society in general. Unfortunately cave restoration may be achievable only over timescales of decades. There is therefore a need for cave managers and scientists to educate the community about the stability and resilience of cave ecosystems and the close interrelationships between surface and underground processes. This can be accomplished by incorporat-ing such information in the interpretation programme of tourist caves, and by raising the public profile of cave use and management.

It is important that clear objectives for the rehabilitation or restoration of caves be established before any work takes place. This necessitates extensive consultation between industry, govern-ment agencies and community interest groups. A primary objective has to be the removal of any sources of pollution, both surface and underground. This may involve sealing of landfills, excavation and removal of contaminated sediments, bioremediation using bacteria or plants, flushing of contaminated water or sediments from caves, or even excavation and removal of contaminants from caves. This is a costly process, and often a portion of the cost will have to be borne by those government agencies responsible for environmental management. This is especially so when the source of pollution is either very old or is community based – for example, sewage pollution or municipal garbage dumps.

Another important objective is the degree to which the original landscape will be simulated, and the final land use to be permitted. In England quarry rehabilitation has involved simulation of the natural geomorphic features of the karst, and revegetation for aesthetic and habitat values (Gunn 1993). The former can be achieved as the final stage in extraction using controlled blasts to create buttress and gully topography, and a variety of regolith textures.

There are a number of basic principles for karst rehabilitation:

- Restore the hydrology of the site. It is crucial to have a high degree of secondary permeability to allow the flow of percolation water and groundwater recharge. This may involve ripping or removal of compacted soil or sediments, construction of drainage control banks, and removal of sediment from infilled dolines or joints.
- Control active soil erosion and sediment entry to the karst system. This may involve construction of contour banks, revegetation, filtration of drainage water at streamsinks, and stabilization of steep slopes prone to mass movement.
- Get the soil biology working again. Invertebrates such as earthworms, ants and termites are effective at breaking down leaf litter, bioturbating the regolith, improving soil texture, increasing the quantity and availability of nutrients, and providing a food source for higher animals.
- Establish a stable vegetation cover, preferably of perennial plants. Permanent vegetation is effective at controlling

erosion by water or wind, provides a source of slow-release nutrients, and is aesthetically pleasing.

- Monitor progress above and below ground. The success of rehabilitation and compliance with community standards can be gauged only by repeated sampling of karst waters. The sampling should be event based, so as to account for increased transfer of sediments and solutes during rainstorms. Ideally monitoring should be continuous, using dataloggers and probes. Relationships between karst water chemistry and suspended load can be developed with simple parameters such as conductivity, pH and turbidity.

- Leave the site alone unless things go wrong. There is a great temptation to interfere with the rehabilitation when processes are slow, especially if political pressure is constant. Revegetation should ideally be formally assessed only after a minimum of two years, when sufficient establishment and growth has occurred. Clearly the site should be regularly examined for excessive sediment transport, and remedial action taken if necessary. For many karst areas, especially those where climate limits biological processes, rehabilitation time may be measured in decades.

The reintroduction of cave fauna is desirable if it is possible. The transfer of organisms from non-impacted sites may be achievable and, with improvements in captive breeding technology, seed populations may be introduced. But for many sites the role of biological refugia may be crucial. In Hidden River Cave, Kentucky, sections of the cave severely polluted by urban sewage have now been recolonized by crustaceans and other invertebrates which survived in tributary passages that were less polluted. While sewage flowed into the cave from the towns of Cave City and Horse Cave, the underground river was dead, populated only by sewage fungi and bloodworms. Toxic fumes produced nausea in the few cavers braving the polluted cave. The sewage was diverted in December 1989, and within two years the stream became oxygenated again, supporting troglophilic crayfish feeding on the dying sewage organisms (Lewis 1993a). One year later the crayfish population had declined, but troglobitic isopods were present in small numbers. These organisms had presumably recolonized the site from small, less polluted fissures in tributary caves. It is therefore very important that such vital refugia remain undisturbed, through limiting access by recreational cavers and cave scientists.

International Cooperation and Liaison

Human visitors to caves may have a significant cumulative impact upon physical and biological values at both the site and the regional level (Spate and Hamilton-Smith 1993). There is therefore a need to prepare and implement management plans that provide access to caves, ensure appropriate limits on visitor numbers where necessary, and institute both minimal impact visitor practices and suitable tracks or other means to protect the environment. There is a range of levels at which international cooperation and liaison may be of considerable assistance. At the simplest, exchange of information may well further the work of those concerned to ensure protection of natural resources. This may take place through exchange of publications, use of electronic media, meetings at conferences or seminars, study visits, and doubtless many other means. A set of guidelines has recently been developed for this purpose by the IUCN Commission on National Parks and Protected Areas.

The International Union of Speleology, with its membership network of national speleological organizations, can and does play a particularly important role in fostering such exchange. In particular it provides a forum which brings together both professional scientists and recreational cave explorers and surveyors. Its *Speleological Abstracts (Bulletin Bibliographique Spéléologique)* provides a continually improving access to world literature. Further, its documentation commission is developing comparable protocols for cave and karst databases, making extensive use of electronic media in doing so. Other scientific organizations, such as the International Geographic Union and the International Association for Hydrogeology, foster and integrate scientific understanding of karst and caves. They also play a fundamentally important role in information exchange. The International Show Caves Association and the newly established International Speleological Heritage Association may also prove valuable, particularly as resources for public education.

The development of expertise within management agencies and the establishment of national or regional bodies, such as the American Cave Conservation Association or the Australasian Cave and Karst Management Association, centrally concerned with cave and karst management or conservation, are also providing an important opportunity to integrate knowledge and understanding as a basis for further dissemination of expertise through information exchange. Those with specialized knowledge and experience in cave and karst protection may well undertake advisory, consultancy or training roles in furthering the protection of cave and karst areas. The various organizations already referred to above provide an avenue for the identification of appropriate experts.

Two or more authorities or even countries may well work under joint-action agreements to share responsibility in protection and management. One well-known example involves the great Aggtelek karst of Central Europe, where close cooperation between two national governments has provided for protection and management of an outstanding karst resource. Such an arrangement provides for holistic management of a specific resource and for congruent strategies to be adopted which span national or other boundaries.

On a smaller but much more widespread basis, responsibility for a karst catchment is often divided between two or more different management tenures. The development of total catchment management policies and programmes on a cooperative basis is vital for adequate protection of the resource in these situations. At another level, the establishment of inter-agency partnerships (perhaps best-practice partnerships – a growing trend in park management) can further the capacity of all parties.

One of the areas in which international information exchange is important is the development of protected area policies. Although these are often generalized and do not deal specifically with particular kinds of resource (e.g., karst), an increasing number of management agencies do have documented karst management policies. There may also be well-recognized policies and practices in place which have not been made explicit in any formal document. Although many cave and karst areas have been included in protected areas (e.g., in the USA, Australia, New Zealand, Thailand, Malaysia and Japan), there is no systematic documentation of which areas are protected in this way. There is also a need to identify major unprotected areas which deserve recognition. An appropriate database, perhaps at the World Conservation Monitoring Centre, should be established at an early date.

A number of cave and karst sites have been recognized under the World Heritage Convention (such as Mammoth Cave, Kentucky) but this may have been for reasons other than their status as karst sites. A review of existing recognized sites should be undertaken in order to a) clarify the application of the heritage criteria to karst sites, and b) identify high-value sites not yet included so that the respective governments might be encouraged to nominate them. International cooperation can play a vital role in strengthening the karst protection and management capacity of land management agencies and in ensuring integrated protection for caves on a world basis.

Restoring a Limestone Ecosystem in Tasmania's World Heritage Area

Quarry rehabilitation is a rapidly growing field in Australia, and at present three karst sites are being treated: Mount Etna in central Queensland, Wombeyan marble quarry in southern New South Wales, and Lune River quarry in the Tasmanian Wilderness World Heritage Area. The Lune River quarry overlies the Exit Cave system, an extensive and complex cave whose geomorphic and faunal values caused it to be inscribed on the World Heritage listing in 1989. The Exit Cave system has 26 km of mapped passages (see figure 10.1) and has formed along three main genetic axes.

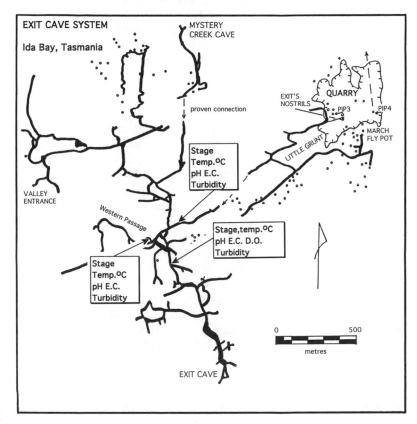

Figure 10.1
*Plan of Exit Cave,
Tasmania and overlying
Lune River quarry
showing location of water
monitoring sites and
parameters continuously
recorded.*

Speleothems on high level sediments in Hammer Passage exceed
the limits of U-series dating at *c.*350 000 years BP. The cave
has been affected by glacial diversion of drainage waters on at least
two occasions (Goede 1969), and it has captured part of both
the D'Entrecasteaux and Lune River catchments in the past. There
is extensive Devonian palaeokarst with sulphide mineralization
that has had a significant effect on karst processes. The cave is
home to a rich terrestrial and aquatic fauna adapted to the
continual dark and cold. In several places the walls are covered
with awesome displays of glow-worms which simulate the night
sky.

A number of impacts have been observed in Exit Cave and its
tributary caves as a result of the quarry operations (Houshold
1992):

- removal of cave passages and their contents by
  quarrying
- destruction of palaeokarstic fills by quarrying
- increased sedimentation of fine clays in Little Grunt Cave
  (underlying the quarry) and the hydrologically connected
  Eastern Passage of Exit Cave
- recurrent turbidity in Eastern Passage and Exit Creek

- changes in pH, conductivity and sulphate ion concentrations (to $150 \, mg \, L^{-1}$) in passages draining the quarry
- re-solution of speleothems by acidified drainage waters because of oxidation of sulphides from palaeokarst fills
- reduced densities of hydrobiid molluscs in passages draining the quarry.

Because the continued operation of the quarry was producing these major geomorphic, water quality and biological impacts on the cave, the quarry was closed in August 1992 under World Heritage legislation. Following the preparation of a rehabilitation plan, a joint Commonwealth–Tasmanian team started active rehabilitation in April 1993. The challenge was then to rehabilitate the quarry and the affected parts of the cave without further impacting the karst values and ecosystems. The primary objective of the rehabilitation has been to protect the World Heritage values of the Exit Cave system by returning the ecosystem processes within the quarry area as close as possible to their original state. The main issues for rehabilitation are the integrity of the underground drainage, its water quality, and the cave invertebrate populations. A secondary objective has been to maintain a high degree of interconnected secondary porosity in the quarry for effective recharge and to simulate as much as possible the original polygonal karst drainage and its forest cover.

The key concept in the rehabilitation is the simulation of the high secondary porosity of a polygonal karst network at a range of spatial scales from that of the whole Exit Cave system down to the diffuse infiltration points within an area of $100 \, m^2$. To achieve these objectives the quarry was subdivided into a number of small closed drainage basins (0.1 to 0.2 ha), each of which has a karst

Plate 10.9
Initial work at the Lune River quarry entailed division of each bench into small closed drainage basins or bunds, and construction of filters around active streamsinks. Topsoil mounds and mulch were then placed in the bunds.

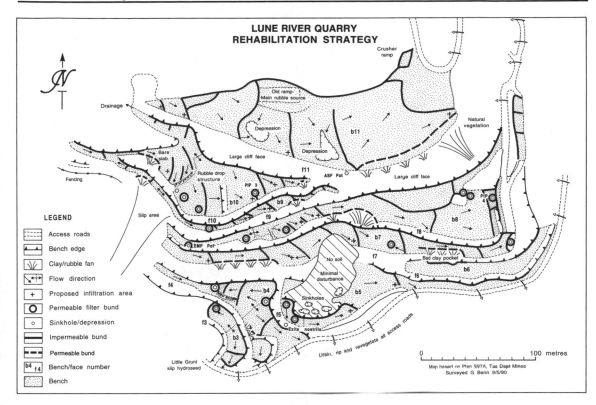

Figure 10.2
Lune River Quarry rehabilitation strategy. From Gillieson and Houshold, in press.

sink or infiltration zone (see figure 10.2 and plate 10.9). Each sink is protected by a filter structure, and areas under clay fans have additional structures to limit the movement of clay after rain. Drainage control has been achieved by several methods. Impermeable bunds of clay and rubble have been constructed along the outer edge of the benches. Similar bund walls have been placed to subdivide the benches into 0.1 to 0.2 ha internal drainage basins which simulate the size of the depressions of polygonal karst (see figure 10.3) on adjoining Marble Hill. This simulation of natural karst drainage should be adequate to restore the dispersed nature of water flow into the Exit Cave hydrological system. Streamsink filters employed limestone blocks, a geotextile filter, crushed rock (20 mm) and straw/mulch in order away from the sink point (see figure 10.4 and plate 10.10). Several additional water sink points became apparent when the highly compacted benches were ripped. On most benches open cavities were present up to 80 m deep. These demonstrate the reality of open, direct hydrologic connections into the Exit Cave system over most of the quarry.

The stabilization of clay fans and cave fills on the benches has been achieved in several ways. First, the diversion of drainage from above by impermeable bunds has reduced gullying by overspillage. Where benches sloped steeply outwards, diffuse infiltration areas of

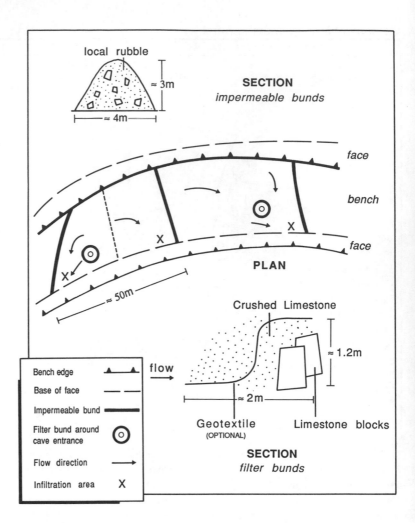

local rubble

≈ 3m

≈ 4m

SECTION
impermeable bunds

face

bench

face

X

X

X

PLAN

≈ 50m

Crushed Limestone

≈ 1.2m

≈ 2m

Bench edge	▲ ▲
Base of face	– – –
Impermeable bund	▬▬▬
Filter bund around cave entrance	⊙
Flow direction	→
Infiltration area	X

flow

Geotextile
(OPTIONAL)

Limestone blocks

SECTION
filter bunds

Figure 10.3
Layout and construction details of small catchment areas or bunds, Lune River Quarry rehabilitation. From Gillieson and Houshold, in press.

Plate 10.10
A filter bund around a streamsink after revegetation. Sedges, leguminous shrubs and eucalypt trees were all used in replanting.

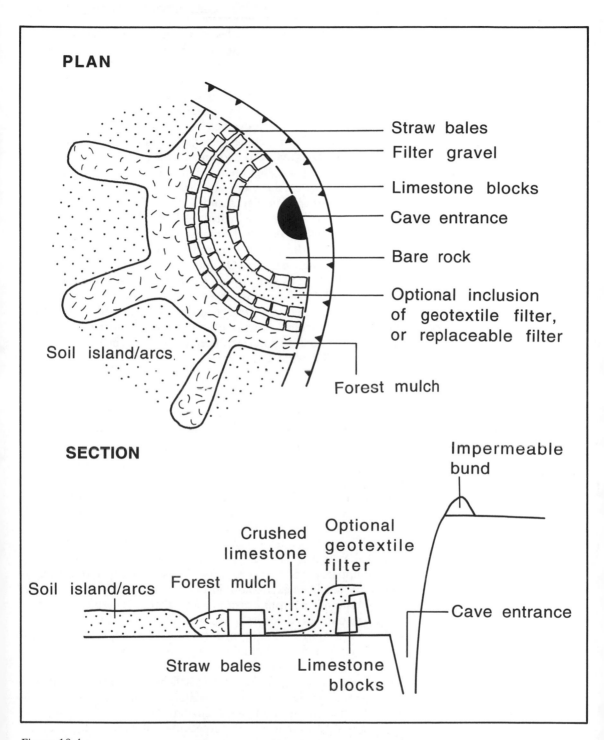

PLAN

Straw bales
Filter gravel
Limestone blocks
Cave entrance
Bare rock
Optional inclusion of geotextile filter, or replaceable filter
Forest mulch
Soil island/arcs

SECTION

Impermeable bund
Crushed limestone
Optional geotextile filter
Forest mulch
Soil island/arcs
Cave entrance
Straw bales
Limestone blocks

Figure 10.4
Layout and construction details of treatment of stream sinks and infiltration areas, Lune River Quarry rehabilitation. From Gillieson and Houshold, in press.

mulch, gravel and straw bales have been constructed to slow water draining over the edge of a face into a gully. Secondly, hydromulching of steep clay fans has probably reduced the risk of rainsplash erosion liberating fine clays. This hydromulch is very thin (less than 5 mm) but forms a stable, if not continuous, cover. Thirdly, the construction of permeable filter arcs below the clay fans has restricted the movement of fines onto the soil-covered benches. The hydromulching was carried out using chopped ink-free newspaper, native seed and an appropriate low fertilizer level. The final coating was between 2 and 5 mm thick and effectively retarded erosion of the clays and loose rock faces over the wet Tasmanian winter. The Parks and Wildlife Service chose to use a slow-release fertilizer in the hydromulch at a level of application of 250 kg ha^{-1}. This was applied to an estimated 0.9 ha of steep faces and clay fans in the quarry, where rapid vegetation establishment was seen as a high priority. On most benches a depth of between 200 and 300 mm of sandy topsoil has been spread and mounded to give numerous small hollows and mounds for detention of rainwater. Numerous roots are present in this material and also a minor source of seed from branches of tea-tree (*Leptospermum* sp.) has been incorporated during clearance operations. The species used in the revegetation were as follows:

Acacia melanoxylon	*Acacia verticillata*
Bedfordia salicina	*Cassinia aculeata*
Eucalyptus nitida	*Eucalyptus obliqua*
Gahnia grandis	*Leptospermum scoparia*
Melaleuca squamea.	

Only limited dye-tracing experiments at the site have been carried out so far (Kiernan 1993). All are low-flow traces using Rhodamine WT. IB47 National Gallery Cave and PIP3 sink have fast positive connections to the Eastern Passage of Exit Cave. PIP4 sink has a fast positive connection to nearby Bradley Chesterman Cave. Flow directions at high flow are unknown and reactivation of fossil passages is quite possible, involving drainage capture in the upper parts of the quarry. Water quality monitoring in Exit Cave has been carried out in two ways, by spot water samples and automated sampling using dataloggers. In addition to field measurements of discharge, temperature, pH and conductivity, samples are being taken for turbidity, total nitrate, total phosphate and major cations and anions. This will give a good set of data from which to evaluate the effectiveness of the rehabilitation programme. The sites in Exit Cave will be maintained for at least two years. The data will allow evaluation of the success of the rehabilitation programme in reducing the inflow of suspended sediments and excess anions (especially sulphates) into the Exit Cave system, as well as the restoration of natural percolation rates into the karst drainage system through analysis of recession hydrographs.

References

Aley, T. 1994: Some thoughts on environmental management as related to cave use. *ACKMA Journal* 17, 4–10.

Bureau of Land Management 1987: *Cave Resources Management*. Washington, DC: US Department of the Interior, 12 pp.

Campbell, B. L., Elliot, G. L. and Loughran, R. J. 1986: Measurement of soil erosion from fallout Caesium-137. *Search* 17, 148–9.

Chapman, P. 1993: *Caves and Cave Life*. London: Harper Collins, 219 pp.

Costin, A. B., Greenaway, M. A. and Wright, L. G. 1984: *Harvesting Water from Land*. Canberra: ANU Centre for Resource and Environmental Studies, 92 pp.

Davey, A. G. 1977: Evaluation criteria for the cave and karst heritage of Australia. *Helictite* 15, 1–41.

Davies, S. J. and Gillieson, D. 1986: The Limestone Flora at Bendethera, Deua, Wadbilliga National Park. Unpublished report to NSW NPWS, Southeastern Region, 15 pp.

Forest Service 1986: *Forest Service Manual: Directive 2356, Cave Management*. Washington, DC: US Forest Service, 6 pp.

Forestry Commission, Tasmania 1987: *Forest Practices Code*. Hobart: Government Printer, 46 pp.

Gams, I., Nicod, J., Sauro, U., Julian, E. and Anthony, U. 1993: Environmental change and human impacts on the Mediterranean karsts of France, Italy and the Dinaric region. In P. Williams (ed.) *Karst Terrains: Environmental Changes and Human Impact*: Catena Supplement 25, 59–98.

Gill, D. 1993: Guidelines for caving and research expeditions to the Mulu National Park, Sarawak. *International Caver* 8, 41.

Gillieson, D. S. 1980: Pit structures from Selminum Tem Cave, Western Province, Papua New Guinea. *Aust. Archaeol.* 10, 26–32.

Gillieson, D. 1989: Limestone soils in the New Guinea Highlands: a review. In D. Gillieson and D. Ingle Smith *Resource Management in Limestone Landscapes: International Perspectives*. Special Publication 2, Department of Geography and Oceanography, University College, Australian Defence Force Academy, 191–200.

Gillieson, D. and Houshold, I. (1996) Rehabilitation of the Lune River quarry, Tasmanian Wilderness World Heritage Area, Australia. In *Karst Hydrogeology and Human Activities: Impacts, Consequences and Implications*. International Association for Hydrogeology.

Gillieson, D., Cochrane, A. and Murray, A. 1994: Surface hydrology and soil erosion in an arid karst: the Nullarbor Plain, Australia. *Environmental Geology* 23, 141–7.

Gillieson, D., Gorecki, P., Head, J. and Hope, G. 1986: Soil Erosion and Agricultural History in the Central Highlands of New Guinea. In V. Gardiner (ed.) *International Geomorphology*. London: John Wiley and Sons, 507–522.

Glasser, N. F. and Barber, G. 1995: Cave conservation plans: the role of English Nature. *Cave and Karst Science* 21, 33–6.

Goede, A. 1969: Underground stream capture at Ida Bay, Tasmania, and the relevance of cold climate conditions. *Aust. Geogr. Stud.* 7, 41–8.

Goldie, H. 1993: Human impact on karst in the British Isles. In P. Williams (ed.) *Karst Terrains: Environmental Changes and Human Impact*: Catena Supplement 25, 161–86 .

Gunn, J. 1993: The geomorphological impacts of limestone quarrying. In P. Williams (ed.) *Karst Terrains: Environmental Changes and Human Impact: Catena Supplement* 25, 187–98.

Hardwick, P. and Gunn, J. 1993: The impact of agriculture on limestone caves. In P. Williams (ed.) *Karst Terrains: Environmental Changes and Human Impact: Catena Supplement* 25, 235–50.

Head, L. 1989: Prehistoric Aboriginal impacts on Australian vegetation: an assessment of the evidence. *Aust. Geog.* 20, 37–46.

Holland, E. 1994: The effects of fire on soluble rock landscapes. *Helictite* 32(1), 3–9.

Houshold, I. 1992: *Geomorphology, Water Quality and Cave Sediments in the Eastern Passage of Exit Cave and its Tributaries*. Report to World Heritage Planning Team, Department of Parks, Wildlife and Heritage, Tasmania, 18 pp.

Kiernan, K. 1988: *The Management of Soluble Rock Landscapes: an Australian Perspective*. Sydney: Speleological Research Council, 61 pp.

Kiernan, K. 1993: The Exit Cave quarry: tracing waterflows and resource policy evolution. *Helictite* 31, 27–42.

Kranjc, A. 1979: The influence of man on cave sedimentation. *Actes Symp. Intern. Erosion Karstique, Aix–Marseille–Nîmes*, 117–23.

Lewis, J. 1993a: The rebirth of Hidden River. *American Caves* 6, 15.

Meiman, J. 1991: The effects of recharge basin land-use practices on water quality at Mammoth Cave National Park, Kentucky. *Proceedings 1991 Cave Management Symposium*, Bowling Green, Kentucky, American Cave Conservation Association, 105–15.

Murray, A. S., Caitcheon, G., Olley, J. and Crockford, H. 1990: Methods for determining the sources of sediments reaching reservoirs: targeting soil conservation, Australian National Committee on Large Dams. *ANCOLD Bull.* 85, 61–70.

National Parks and Wildlife Service 1983: *Harvesting and Rehabilitation of Jounama Pine Plantation, Kosciusko National Park: Environmental Impact Statement*. NPWS (NSW) and Forestry Commission (NSW), 198 pp.

Northcote, K. H. 1979: *A Factual Key for the Recognition of Australian Soils*, 4th edn. Adelaide: Rellim Technical Publications.

O'Neill, A. L., Head, L. M. and Marthick, J. K. 1993: Integrating remote sensing and spatial analysis techniques to compare Aboriginal and pastoral fire patterns in the East Kimberley, Australia. *Applied Geography* 13, 67–85.

Rosewell, C. J. and Edwards, K. 1988: SOILOSS: a program to assist in the selection of management practices to reduce erosion. *Soil Conservation Service of NSW Technical Handbook* no. 11. Sydney, 71 pp plus disk.

Spate, A. and Hamilton-Smith, E. 1993: Caver's impacts – some theoretical and applied considerations. *Proc. 9th Australasian Cave Tourism and Management Conference*, Margaret River, WA, 20–30.

Stanton, R. K., Murray, A. S. and Olley, J. M. 1992: Tracing recent sediment using environmental radionuclides and mineral magnetics in the karst of Jenolan Caves, Australia. In J. Bogen, D. E. Walling and T. J. Day (ed.) *Erosion and Sediment Transport Monitoring Programmes in River Basins*. Proceedings of a symposium held at Oslo, August 1992. IAHS Publ. no. 210. Wallingford, UK: IAHS Press.

Urich, P. 1991: Stress on tropical karst resources exploited for the cultivation of wet rice. In U. Sauro, A. Bondesan and M. Meneghel (ed.)

Proceedings of the International Conference on Environmental Changes in Karst Areas. Italy, 15–27 Sept. 1991. Quad. Dip. Geografia, Università di Padova, 39–48.

Urushibara-Yoshino, K. 1993: Human impact on karst soils: Japanese and other examples. In P. Williams (ed.) *Karst Terrains: Environmental Changes and Human Impact*: *Catena Supplement* 25, 219–34.

Walling, D. E. and Quine, T. A. 1990: Calibration of Caesium-137 measurements to provide quantitative erosion rate data. *Land Degradation and Rehabilitation* 2, 161–75.

Waltham, A. C. 1983: A review of karst conservation sites in Britain. *Studies in Speleology* 4, 85–92.

White, W. B. 1988: *Geomorphology and Hyorology of Karst Terrains*. New York: Oxford University Press, 464 pp.

Williams, J. E., Whelan, R. J. and Gill, A. M. 1994: Fire and environmental heterogeneity in southern temperate forest ecosystems: implications for management. *Aust. J. Botany* 42, 125–37.

Young, E. A. 1992: Aboriginal land in Australia: expectations, achievements and implications. *Applied Geography* 12, 146–61.

Yuan, D. 1993: Environmental change and human impact on karst in southern China. In P. Williams (ed.) *Karst Terrains: Environmental Changes and Human Impact*: *Catena Supplement* 25, 99–108.

Further Reading

Print Media

The following books may provide the curious with a greater depth of understanding of caves and karst. In particular, the excellent volume by Ford and Williams (1989) has become the single most referenced and comprehensive book on karst and caves in the English language.

Chapman, P. 1993: *Caves and Cave Life*. London: Harper Collins, 219 pp.

Courbon, P., Chabert, C., Bosted, P. and Lindsley, K. 1989: *Atlas of the Great Caves of the World*. St Louis: Cave Books, 368 pp.

Culver, D. C. 1982: *Cave Life: Evolution and Ecology*. Cambridge, MA: Harvard University Press, 189 pp.

Dreybrodt, W. 1988: *Processes in Karst Systems: Physics, Chemistry, and Geology*. Berlin: Springer, 288 pp.

Ford, D. C. and Williams, P. W. 1989: *Karst Geomorphology and Hydrology*. London: Unwin Hyman, 601 pp.

Jennings, J. N. 1985: *Karst Geomorphology*. Oxford: Blackwell, 293 pp.

Shaw, T. R. 1992: *History of Cave Science*. Sydney: Speleological Research Council, 338 pp.

Trudgill, S. T. 1985: *Limestone Geomorphology*. London: Longmans, 196 pp.

White, W. B. 1988: *Geomorphology and Hydrology of Karst Terrains*. New York: Oxford University Press, 464 pp.

Electronic Media Sources

There are now several sites on the Internet which provide a wide range of information on speleology and cave science. Several of these have graphic material in the form of maps and photographs, and have active links to related servers, providing information on books, government reports and equipment. In addition there is an electronic mail server, the Cavers Digest, which provides an eclectic mixture of caving news, conservation action and opinions.

International Union of Speleology (Informatics Commission)
   http://rubens.its.unimelb.edu.au/~pgm/uis/

International Subterranean Heritage Association (Belgium)
    http://www.microsearch.be/isha/
Sherry Mayo's Cave Pages (Australia)
    http://rschp2.anu.edu.au:8080/cave/cave.html
Western Australian Speleology Group (Australia)
    http://techpkwa.curtin.edu.au/interests/Speleology/intro.html
The Speleonuts (Brazil)
    http://www.prodemge.gov.br/~marcosf/speleonuts.html
Section INRIA de Spéléologie Caving Server (France)
    http://www.inria.fr/agos-sophia/sis/sis.html
Cornelia Klumper's Munich Speleo Server (Germany)
    http://bigbang.usm.uni-muenchen.de:8002/~conny/cave/server.html
Waitomo Caves and Museum (New Zealand)
    http://www3.waikato.ac.nz/waitomo/
Lancaster University Speleological Society (UK)
    http://www.comp.lancs.ac.uk/rec-and-travel/luss/index.html
Andrew Brook's Cave Server (UK)
    http://www.sat.dundee.ac.uk/~arb/speleo.html
The Speleology Home Page (USA; also has Cavers' Digest email archive)
    http://speleology.cs.yale.edu/pub/caving/
Paul Aughey's Caving Page (USA)
    http://freenet3.scri.fsu.edu:81/users/caver/html/cavepage.html
Central Kentucky Karst Network (USA)
    http://www.wku.edu/~glennja/karst.html
NSS Cave Conservation and Management Section Page (USA)
    http://www.halcyon.com/samara/nssccms/welcome.html

Glossary of Cave and Karst Terminology

Modified and extended from that compiled by J. N. Jennings for the Australian Speleological Federation

Abbreviations and conventions
Abb. = abbreviation
Syn. = synonym (word with same meaning)
Cf. = confer (compare) with the following term which is not identical but related to it
n. = noun
v. = verb

ACCIDENTAL (n.) An animal accidentally living in a cave.

ACTIVE CAVE A cave which has a stream flowing in it.

ADAPTATION An inherited characteristic of an organism in structure, function or behaviour which makes it better able to survive and reproduce in a particular environment. Lengthening of appendages, loss of pigment and modification of eyes are considered adaptations to the dark zone of caves.

AEOLIAN CALCARENITE A limestone formed on land by solution and redeposition of calcium carbonate in coastal dune sands containing a large proportion of calcareous sand from mollusc shells and other organic remains.

AGGRESSIVE Referring to water which is still capable of dissolving more limestone, other karst rock, or speleothems.

ALLOGENIC Referring to water or sediment which has a source on non-karstic rocks.

ANASTOMOSIS A mesh of tubes or half-tubes.

ARAGONITE A less common crystalline form of calcium carbonate than calcite, denser and orthorhombic.

ARTHROPODS The most common group of animals inhabiting caves, including insects, crustaceans, spiders, millipedes, etc. They have jointed limbs and external skeletons.

AUTOGENIC Referring to water or sediment which is derived from karstic rocks.

AUTOTROPH A green plant, bacterium or protist which manufactures complex organic compounds (food) from simple inorganic raw materials, using a source of energy from light or chemical compounds.

BARE KARST Karst with much exposed bedrock.

BAT A member of the order Chiroptera, the only mammals capable of true flight as they have membranes between the toes of their forefeet.

BATHYPHREATIC Referring to water moving with some speed through downward looping passages in the phreatic zone.

BED A depositional layer of sedimentary bedrock or unconsolidated sediment.

BEDDING-GRIKE A narrow, rectilinear slot in a karst rock outcrop as a result of solution along a bedding-plane.

BEDDING-PLANE A surface separating two beds, usually planar.

BEDDING-PLANE CAVE A cavity developed along a bedding-plane and elongate in cross-section as a result.

BENTHIC Bottom dwelling, usually on the bed of a stream, pond, lake or the sea.

BIOMASS The total weight of living matter in a given area, or in a community, at a particular trophic level, or of a particular type of organism at a site.

BIOSPELEOLOGY The scientific study of organisms living in caves.

BLIND SHAFT A vertical extension upwards from part of a cave but not reaching the surface; small in area in relation to its height.

BLIND VALLEY A valley that is closed abruptly at its lower end by a cliff or slope facing up the valley. It may have a perennial or intermittent stream which sinks at its lower end or it may be a dry valley.

BLOWHOLE (1) A hole to the surface in the roof of a sea cave through which waves force air and water. (2) A hole in the ground through which air blows in and out strongly, sometimes audibly; common in the Nullarbor Plain.

BONE BRECCIA A breccia containing many bone fragments.

BRANCHWORK A dendritic system of underground streams or passages wherein branches join successively to form a major stream or passage.

BREAKDOWN Fall of rock from the roof or wall of a cave.

BRECCIA Angular fragments of rock and/or fossils cemented together or with a matrix of finer sediment. Cf. Bone breccia.

CALCITE The commonest calcium carbonate ($CaCO_3$) mineral and the main constituent of limestone, with different crystal forms in the rhombohedral subsystem.

CANOPY A compound speleothem consisting of a flowstone cover of a bedrock projection and of a fringe of stalactites or shawls on the outer edge.

CANYON (1) A deep valley with steep to vertical walls; in karst frequently formed by a river rising on impervious rocks outside the karst area. (2) A deep, elongated cavity cut by running water in the roof or floor of a cave or forming a cave passage.

CARBIDE Calcium carbide, CaC_2, used with water to make acetylene in lamps.

CAVE A natural cavity in rock large enough to be entered by man. It may be water-filled. If it becomes full of ice or sediment and is impenetrable, the term applies but will need qualification.

CAVE BLISTER An almost perfect hemisphere of egg-shell calcite.

CAVE BREATHING (1) Movement of air in and out of a cave entrance at intervals. (2) The associated air currents within the cave.

CAVE CORAL Very small speleothems consisting of short stalks with bulbous ends, usually occurring in numbers in patches.

CAVE EARTH Clay, silt, fine sand and/or humus deposited in a cave.

CAVE ECOLOGY The study of the interaction between cave organisms and their environment, e.g., energy input from surface, climatic influences.

CAVE FILL Transported materials such as silt, clay, sand and gravel which cover the bedrock floor or partially or wholly block some part of a cave.

CAVE FLOWER Syn. gypsum flower.

CAVE PEARL A smooth, polished and rounded speleothem found in shallow hollows into which water drips. Internally it has concentric layers around a nucleus.

CAVE SPRING A natural flow of water from rock or sediment inside a cave.

CAVE SYSTEM A collection of caves interconnected by enterable passages or linked hydrologically or a cave with an extensive complex of chambers and passages.

CAVERN A very large chamber within a cave.

CAVERNICOLE An animal which normally lives in caves for the whole or part of its life cycle.

CAVING The entering and exploration of caves.

CENOTE A partly water-filled, wall-sided doline.

CHAMBER The largest order of cavity in a cave, with considerable width and length but not necessarily great height.

CHERT A light grey to black or red rock, which fractures irregularly, composed of extremely fine crystalline silica and often occurring as nodules or layers in limestone.

CHOKE Rock debris or cave fill blocking part of a cave.

COLUMN A speleothem from floor to ceiling, formed by the growth and joining of a stalactite and a stalagmite, or by the growth of either to meet bedrock.

CONDUIT An underground stream course completely filled with water and under hydrostatic pressure or a circular or elliptical passage inferred to have been such a stream course.

COPROLITE Fossilized large excrement of animals, sometimes found in caves, especially those used as lairs.

COPROPHAGE A scavenger which feeds on animal dung, including guano.

CORRASION The wearing away of bedrock or loose sediment by mechanical action of moving agents, especially water.

CORROSION Syn. solution.

CRAWL (WAY) A passage which must be negotiated on hands and knees.

CROSS-SECTION A section of a cave passage or a chamber across its width.

CRYPTOZOA The assemblage of small terrestrial animals found living in darkness beneath stones, logs, bark, etc. Potential colonizers of caves.

CRYSTAL POOL A cave pool generally with little or no overflow, containing well-formed crystals.

CURRENT MARKING Shallow asymmetrical hollows formed by solution by turbulent waterflow and distributed regularly over karst rock surfaces. Cf. Scallops.

CURTAIN A speleothem in the form of a wavy or folded sheet, often translucent and resonant, hanging from the roof or wall of a cave.

DARK ADAPTATION A change in the retina of the eye sensitizing it to dim light (the eye 'becomes accustomed to the dark'). Loss of sensitivity on re-exposure to brighter light is 'light adaptation'.

DARK ZONE The part of a cave which daylight does not reach.

DAYLIGHT HOLE An opening to the surface in the roof of a cave.

DEAD CAVE A cave without streams or drips of water.

DECOMPOSERS Living things, chiefly bacteria and fungi, that subsist by extracting energy from tissues of dead animals and plants.

DECORATION Cave features as a result of secondary mineral precipitation, usually of calcite. Syn. Speleothem.

DETRITIVORE Organisms which feed on organic detritus, such as the dead parts of plants or the dead bodies and waste products of animals.

DIG An excavation made (1) to discover or extend a cave or (2) to uncover artefacts or animal bones.

DIP The angle at which beds are inclined from the horizontal. The true dip is the maximum angle of the bedding-planes at right angles to the strike. Lesser angles in other directions are apparent dips.

DOG-TOOTH SPAR A variety of calcite with acute-pointed crystals.

DOLINE A closed depression draining underground in karst, of simple but variable form, e.g., cylindrical, conical, bowl- or dish-shaped. From a few to many hundreds of metres in dimension.

DOLOMITE (1) A mineral consisting of the double carbonate of magnesium and calcium $CaMg(CO_3)_2$. (2) A rock made chiefly of dolomite mineral.

DOME A large hemispheroidal hollow in the roof of a cave, formed by breakdown and/or salt weathering, generally in mechanically weak rocks, which prevents bedding and joints dominating the form.

DONGA In the Nullarbor Plain, a shallow, closed depression, several metres deep and hundreds of metres across, with a flat clay-loam floor and very gentle slopes.

DRIPHOLE A hole formed by water dripping onto the cave floor.

DRIPLINE A line on the ground at a cave entrance formed by drips from the rock above. Useful in cave survey to define the beginning of the cave.

DRIPSTONE A deposit formed from drops falling from cave roofs or walls, usually of calcite.

DRY CAVE A cave without a running stream. Cf. Dead cave.

DRY VALLEY A valley without a surface stream channel.

DUCK(-UNDER) A place where water is at or close to the cave roof for a short distance so that it can be passed only by submersion.

DUNE LIMESTONE Syn. Aeolian calcarenite.

DYE GAUGING Determining stream discharge by inserting a known quantity of dye and measuring its concentration after mixing.

DYNAMIC PHREAS A phreatic zone or part of a phreatic zone where water moves fast with turbulence under hydrostatic pressure.

ECCENTRIC A speleothem of abnormal shape or attitude. Cf. Helictite.

EPIPHREATIC Referring to water moving with some speed in the top of the phreatic zone or becoming part of the phreatic zone during floods.

EROSION The wearing away of bedrock or sediment at the surface or in caves by mechanical and chemical actions of all moving agents, such as rivers, wind and glaciers.

EXSURGENCE A spring fed only by percolation water.

FAULT A fracture separating two parts of a once continuous rock body with relative movement along the fault plane.

FAULT CAVE A cave developed along a fault or fault zone, either by movement of the fault or by preferential solution along it.

FAULT PLANE A plane along which movement of a fault has taken place.

FISSURE An open crack in rock or soil.

FISSURE CAVE A narrow, vertical cave passage, often developed along a joint but not necessarily so. Usually a result of solution but sometimes of tension.

FLATTENER A passage, which, though wide, is so low that movement is possible only in a prone position.

FLOE CALCITE Very thin flakes of calcite floating on the surface of a cave pool or previously formed in this way.

FLOWSTONE A deposit formed from thin films or trickles of water over floors or walls, usually of calcite. Cf. Travertine.

FLUORESCEIN A reddish-yellow organic dye which gives a green fluorescence to water. It is detectable in very dilute solutions, so it is used in water tracing and dye gauging in the form of the salt sodium fluorescein.

FLUOROMETER An instrument for measuring the fluorescence of water; used in water tracing and dye gauging.

FOSSIL The remains or traces of animals or plants preserved in rocks or sediments.

GLACIER CAVE A cave formed within or beneath a glacier.

GOUR Syn. Rimstone dam.

GRIKE A deep, narrow, vertical or steeply inclined rectilinear slot in a rock outcrop as a result of solution along a joint.

GROTTO A room in a cave of moderate dimensions but richly decorated.

GROUNDWATER Syn. Phreatic water.

GUANO Large accumulations of dung, often partly mineralized, including rock fragments, animal skeletal material and products of reactions between excretions and rock. In caves, it is derived from bats and to a lesser extent from birds.

GYPSUM The mineral hydrated calcium sulphate $CaSO_4.2H_2O$.

GYPSUM FLOWER An elongated and curving deposit of gypsum on a cave surface.

HALF-BLIND VALLEY A blind valley which overflows its threshold when the streamsink cannot accept all the water at a time of flood.

HALF-TUBE A semi-cylindrical, elongate recess in a cave surface, often meandering or anastomosing.

HALITE The sodium chloride mineral NaCl in the cubic crystalline system.

HALL A lofty chamber considerably longer than it is wide.

HELICTITE A speleothem, which at one or more stages of its growth changes its axis from the vertical to give a curving or angular form.

HETEROTROPH An organism that cannot manufacture food from inorganic raw materials, and therefore must feed on available organic compounds contained in the tissues of other organisms.

HISTOPLASMOSIS A lung disease which may be caught from the guano of some caves, caused by a fungus, Histoplasmosis capsulatum. Usually mild in effect, it can be fatal in rare cases.

HYDROLOGY The scientific study of the nature, distribution and behaviour of water.

HYDROSTATIC PRESSURE The pressure due to a column of water.

ICE CAVE A cave with perennial ice in it.

INFLOW CAVE A cave into which a stream enters or is known to have entered formerly but which cannot be followed downstream to the surface.

INTERSTITIAL MEDIUM Spaces between grains of sand or fine gravel filled with water which may contain organisms.

INVERTED SIPHON A siphon of U-profile.

JOINT A planar or gently curving crack separating two parts of once continuous rock without relative movement along its plane.

JOINT-PLANE CAVE A cavity developed along a joint and elongated in cross-section.

KANKAR (pronounced kunkar) A deposit, often nodular, of calcium carbonate formed in soils of semi-arid regions. Sometimes forms cave roofs.

KARREN The minor forms of karst as a result of solution of rock on the surface or underground.

KARST Terrain with special landforms and drainage characteristics on account of greater solubility of certain rocks in natural waters than is common. Derived from the geographical name of part of Slovenia.

KARST WINDOW A closed depression, not a polje, which has a stream flowing across its bottom.

KEYHOLE (PASSAGE) A small passage or opening in a cave, which is round above and narrow below.

LAKE In caving, a body of standing water in a cave. The term is used for what on the surface would be called a pond or pool.

LAVA-CAVE A cave in a lava flow; usually a tube or tunnel formed by flow of liquid lava through a solidified mass, or by roofing of an open channel of flowing lava. Small caves in lava also form as gas blisters.

LEUCOPHOR A colourless water tracer, which fluoresces blue.

LIMESTONE A sedimentary rock consisting mainly of calcium carbonate $CaCO_3$.

LIVE CAVE A cave containing a stream or active speleothems.

MALADIE VERTE Overgrowths of the green alga *Palmellococcus* on speleothems.

MARBLE Limestone recrystallized and hardened by pressure and heat.

MAZE Syn. Network.

MEANDER An arcuate curve in a river course as a result of a stream eroding sideways.

MEANDER NICHE A hemispherically roofed part of a cave formed by a stream meandering and cutting down at the same time.

MICROCLIMATE The climate (i.e., temperature, humidity, air movements, etc.) of a restricted area or space, e.g., of a cave or, on a lesser scale, of the space beneath stones in a cave.

MICROGOUR Miniature rimstone dams with associated tiny pools of the order of 1 cm wide and deep on flowstone.

MOONMILK A soft, white plastic speleothem consisting of calcite, hydrocalcite, hydromagnesite or huntite.

MUD PENDULITE A pendulite with the knob coated in mud.

NATURAL ARCH An arch of rock formed by weathering.

NATURAL BRIDGE A bridge of rock spanning a ravine or valley and formed by erosive agents.

NECROPHAGE A scavenger feeding on animal carcases.

NETWORK A complex pattern of repeatedly connecting passages in a cave.

NICHE The unique life strategy of a species from an ecological viewpoint.

NOTHEPHREATIC Referring to water moving slowly in cavities in the phreatic zone.

OUTFLOW CAVE A cave from which a stream flows or formerly did so and which cannot be followed upstream to the surface.

PALAEOKARST 'Fossil' karst – cave or karst features remnant from a previous period of karstification, characterized by the presence of ancient (buried) deposits, as lithified cave fills or breccias.

PASSAGE A cavity which is much longer than it is wide or high and may join larger cavities.

PENDANT Syn. Rock pendant.

PENDULITE A kind of stalactite which has been partly submerged and the submerged part covered with dog-tooth spar to give the appearance of a drumstick.

PERCOLATION WATER Water moving mainly downwards through pores, cracks and tight fissures in the vadose zone.

PERMEABILITY The property of rock or soil permitting water to pass through it. Primary permeability depends on interconnecting pores between the grains of the material. Secondary permeability depends on solutional widening of joints and bedding-planes and on other solution cavities in the rock.

PHREAS Syn. Phreatic zone.

PHREATIC WATER Water below the level at which all voids in the rock are completely filled with water.

PHREATIC ZONE Zone where voids in the rock are completely filled with water.

PHYTOPHAGE An animal that feeds on green plants.

PILLAR A bedrock column from roof to floor left by the removal of surrounding rock.

PIPE A tubular cavity projecting as much as several metres down from the surface into karst rocks and often filled with earth, sand, gravel, breccia, etc.

PITCH A vertical or nearly vertical part of a cave for which ladders or ropes are normally used for descent or ascent.

PLAN A plot of the shape and details of a cave projected vertically onto a horizontal plane at a reduced scale.

PLUNGE POOL A swirlhole, generally of large size, occurring at the foot of a waterfall or rapid, on the surface or underground.

POLJE A large closed depression draining underground, with a flat floor across which there may be an intermittent or perennial stream and which may be liable to flood and become a lake. The floor makes a sharp break with parts of surrounding slopes.

POLYGONAL KARST Karst completely pitted by closed depressions so that divides between them form a crudely polygonal network.

POOL DEPOSIT (1) Any sediment which has accumulated in a pool in a cave. (2) Crystalline deposits precipitated in a cave pool, usually of crystalline shape as well as structure.

POPULATION Individuals of a species in a given locality which potentially form a single interbreeding group separated by physical barriers from other such populations (e.g., populations of the same species in two quite separate caves).

POROSITY The property of rock or soil of having small voids between the constituent particles. The voids may not interconnect.

POT(HOLE) A vertical or nearly vertical shaft or chimney open to the surface.

PREDATOR An animal which captures other animals for its food.

PROJECTED SECTION The result of projecting a section composed of several parts with differing directions onto a single plane. Usually the plane is vertical along the general trend of the cave. Only the vertical distance apart of points is correct, not the horizontal, so that slopes are distorted.

PSEUDOKARST Terrain with landforms which resemble those of karst but which are not the product of karst processes.

RELICT KARST Old cave forms produced by earlier geomorphic processes within the present cycle of karstification and open to modification by present-day processes such as deposition of speleothems, sediments or skeletal deposits.

RELICT SPECIES Species belonging to an ancient group whose distribution is now restricted to a few locations and whose population is not increasing.

RESURGENCE A spring where a stream, which has a course on the surface higher up, reappears at the surface.

RHODAMINE A red organic dye which gives a red fluorescence to water. It is detectable in very dilute solutions so it is used in water tracing and dye gauging.

RIFT A long, narrow, high and straight cave passage controlled by planes of weakness in the rock. Cf. Fissure.

RIMSTONE A deposit formed by precipitation from water flowing over the rim of a pool.

RIMSTONE DAM A ridge or rib of rimstone, often curved convexly downstream.

RIMSTONE POOL A pool held up by a rimstone dam.

RISING Syn. Spring.

ROCK PENDANT A smooth-surfaced projection from the roof of a cave as a result of solution. Usually found in groups.

ROCK SHELTER A cave with a more or less level floor reaching only a short way into a hillside or under a fallen block so that no part is beyond daylight.

ROCKHOLE A shallow, small hole in rock outcrops, often rounded in form and holding water after rains. Well known on the Nullarbor Plain.

ROCKMILK Syn. Moonmilk.

ROCKPILE A heap of blocks in a cave, roughly conical or part-conical in shape.

ROOF CRUST Thin speleothem on a cave precipitated from water films exuding from pores or cracks.

ROOM A wider part of a cave than a passage but not as large as a chamber.

SALT WEATHERING Detachment of particles of various sizes from a rock surface by the growth of crystals from salt solutions. Forms substantial features in desert caves.

SAPROPHAGE A scavenger feeding on decaying organic material.

SATURATED (1) Referring to rock with water-filled voids. (2) Referring to water which has dissolved as much limestone or other karst rock as it can under normal conditions.

SCALLOPS Current markings that intersect to form points which are directed downstream.

SCAVENGER An animal that eats dead remains and wastes of other animals and plants (cf. Coprophage, Necrophage, Saprophage).

SEA CAVE A cave in present-day or emerged sea cliffs, formed by wave attack or solution.

SECTION A plot of the shape and details of a cave in a particular intersecting plane called the section plane, which is usually vertical.

SEDIMENT Material recently deposited by water, ice or wind, or precipitated from water.

SEEPAGE WATER Syn. Percolation water.

SELENITE A crystalline form of gypsum.

SHAFT A vertical cavity roughly equal in horizontal dimensions but much deeper than broad.

SHAWL A simple triangular-shaped curtain.

SHOW CAVE A cave that has been made accessible to the public for guided visits.

SIPHON A water-filled passage of inverted U-profile which delivers a flow of water whenever the head of water upstream rises above the top of the inverted U.

SOLUTION In karst studies, the change of bedrock from the solid state to the liquid state by combination with water. In physical solution the ions of the rock go directly into solution without transformation. In chemical solution acids take part, especially the weak acid formed by carbon dioxide (CO_2).

SOLUTION FLUTE A solution hollow running down the maximum slope of the rock, of uniform fingertip width and depth, with sharp ribs between it and its neighbours.

SOLUTION PAN A dish-shaped depression on flattish rock; its sides may overhang and carry solution flutes. Its bottom may have a cover of organic remains, silt, clay or rock fragments.

SOLUTION RUNNEL A solution hollow running down the maximum slope of the rock, larger than a solution flute and increasing in depth and width down its length. Thick ribs between neighbouring runnels may be sharp and carry solution flutes.

SPECIES A group of actually or potentially interbreeding populations which is reproductively isolated from other such groups by their biology, not simply by physical barriers.

SPELEOGEN A cave feature formed erosionally or by weathering in cave enlargement, such as current markings (Cf. Scallop) or rock pendants.

SPELEOLOGY The exploration, description and scientific study of caves and related phenomena.

SPELEOTHEM A secondary mineral deposit formed in caves, most commonly calcite.

SPLASH CUP A shallow cavity in the top of a stalagmite.

SPONGEWORK A complex of irregular, interconnecting cavities intricately perforating the rock. The cavities may range from a few centimetres to more than a metre across.

SPRING A natural flow of water from rock or soil onto the land surface or into a body of surface water.

SQUEEZE An opening in a cave passable only with effort because of its small dimensions. Cf. Flattener, Crawl (way).

STALACTITE A speleothem hanging downwards from a roof or wall, of cylindrical or conical form, usually with a central hollow tube.

STALAGMITE A speleothem projecting vertically upwards from a cave floor and formed by precipitation from drips.

STEEPHEAD A steep-sided valley in karst, generally short, ending abruptly upstream where a stream emerges or formerly did so.

STRAW (STALACTITE) A long, thin-walled tubular stalactite less than about 1 cm in diameter.

STREAMSINK A point at which a surface stream disappears underground.

STRIKE The direction of a horizontal line in a bedding-plane in rocks inclined from the horizontal. On level ground it is the direction of outcrop of inclined beds.

STYLOLITE Suture in rock formed where pressure solution has taken place, often leaving a thin lamina of insoluble material along it.

SUBCUTANEOUS ZONE The uppermost layers of rock below the soil on a karst. This zone is distinguished from lower zones by a higher porosity and storage capacity for water as a result of the presence of many solutionally enlarged fissures.

SUBJACENT KARST Karst developed in soluble beds underlying other rock formations; the surface may or may not be affected by the karst development.

SUMP A point in a cave passage when the water meets the roof.

SUPERSATURATED Referring to water that has more limestone or other karst rock in solution than the maximum corresponding to normal conditions.

SURVEY In caving, the measurement of directions and distances between survey points and of cave details from them, and the plotting of cave plans and sections from these measurements either graphically or after computation of coordinates.

SWIRLHOLE A hole in rock in a streambed eroded by eddying water, with or without sand or pebble tools.

SYNGENETIC KARST Karst developed in Aeolian calcarenite when the evolution of karst features has taken place at the same time as the lithification of dune sand.

TAFONI Roughly hemispherical hollows weathered in rock either at the surface or in caves.

TERRA ROSSA Reddish residual clay soil developed on limestone.

THRESHOLD (1) That part of a cave near the entrance where surface climatic conditions rapidly grade into cave climatic conditions. Not necessarily identical with the twilight zone. (2) Slope or cliff facing up a blind or half-blind valley below a present or former streamsink.

THROUGH CAVE A cave which may be followed from entrance to exit along a stream course or along a passage which formerly carried a stream.

TOWER KARST Conekarst in which the residual hills have very steep to overhanging lower slopes. There my be alluvial plains between the towers and flat-floored depressions within them.

TRACER (1) A material introduced into surface or underground water where it disappears or into soil to determine drainage interconnections and travel time. (2) A material introduced into cave air to determine cave interconnections.

TRAVERTINE Compact calcium carbonate deposit, often banded, precipitated from spring, river or lake water. Cf. Tufa.

TROGLOBITE A cavernicole unable to live outside the cave environment.

TROGLODYTE A human cave dweller.

TROGLOPHILE A cavernicole which frequently completes its life cycle in caves but is not confined to this habitat.

TROGLOXENE A cavernicole which spends only part of its life cycle in caves and returns periodically to the surface for food.

TUBE A cave passage of smooth surface, and elliptical or nearly circular in cross-section.

TUFA Spongy or vesicular calcium carbonate deposited from spring, river or lake waters. Cf. Travertine.

TUNNEL A nearly horizontal cave open at both ends, fairly straight and uniform in cross-section.

TWILIGHT ZONE The part of a cave to which daylight penetrates.

UVALA A complex closed depression with several lesser depressions within its rim.

VADOSE FLOW Water flowing in free-surface streams in caves.

VADOSE SEEPAGE Syn. Percolation water.

VADOSE WATER Water in the vadose zone.

VADOSE ZONE The zone where voids in the rock are partly filled with air and through which water descends under gravity.

VAUCLUSIAN SPRING A spring rising up a deep, steeply inclined, water-filled passage into a small surface pool.

VERMICULATION Pattern of thin, worm-shaped coatings of clay or silt on cave surfaces.

WATER TABLE The surface between phreatic water, which completely fills voids in the rock, and ground air, which partially fills higher voids.

WATER TRACING Determination of water connection between points of stream disappearance or of soil water seepage and points of reappearance on the surface or underground.

WATERTRAP A place where a cave roof dips under water but lifts above it farther on. Cf. Duck(-under).

WELL A deep rounded hole in a cave floor or on the surface in karst.

Index